Sabine Reisinger

Organometallic Pnictogen Chemistry – Three Aspects

# Organometallic Pnictogen Chemistry – Three Aspects

Dissertation zur Erlangung des Doktorgrades der Naturwissenschaften (Dr. rer. nat.)
der Fakultät für Chemie und Pharmazie der Universität Regensburg
vorgelegt von

Sabine Reisinger, geb. Scheuermayer

Regensburg
2014

Die Arbeit wurde von Prof. Dr. Manfred Scheer angeleitet.
Das Promotionsgesuch wurde am 20.06.2014 eingereicht.
Das Kolloquium fand am 11.07.2014 statt.

| Prüfungsausschuss: | Vorsitzender: | Prof. Dr. Helmut Motschmann |
|---|---|---|
| | 1. Gutachter: | Prof. Dr. Manfred Scheer |
| | 2. Gutachter: | Prof. Dr. Henri Brunner |
| | weiterer Prüfer: | Prof. Dr. Bernhard Dick |

**Dissertationsreihe der Fakultät für Chemie und Pharmazie der Universität Regensburg, Band 4**

Herausgegeben vom Alumniverein Chemie der Universität Regensburg e.V.
in Zusammenarbeit mit Prof. Dr. Burkhard König, Prof. Dr. Joachim Wegener,
Prof. Dr. Arno Pfitzner und Prof. Dr. Werner Kunz.

# Organometallic Pnictogen Chemistry
# – Three Aspects

**Sabine Reisinger**

Universitätsverlag Regensburg

Bibliografische Informationen der Deutschen Bibliothek.
Die Deutsche Bibliothek verzeichnet diese Publikation
in der Deutschen Nationalbibliografie. Detailierte bibliografische Daten
sind im Internet über http://dnb.ddb.de abrufbar.

1. Auflage 2014

Leibnizstraße 13, 93055 Regensburg

Umschlagentwurf: Alumniverein Chemie der Universität Regensburg e.V.
Layout: Sabine Reisinger
Druck: Docupoint, Magdeburg
ISBN: 978-3-86845-118-4



Weitere Informationen zum Verlagsprogramm erhalten Sie unter:
www.univerlag-regensburg.de

UR
Universität Regensburg

Eidesstattliche Erklärung

Ich erkläre hiermit an Eides statt, dass ich die vorliegende Arbeit ohne unzulässige Hilfe Dritter und ohne Benutzung anderer als der angegebenen Hilfsmittel angefertigt habe; die aus anderen Quellen direkt oder indirekt übernommenen Daten und Konzepte sind unter Angabe des Literaturzitats gekennzeichnet.

____________________________

Sabine Reisinger

This thesis was elaborated within the period from December 2009 till June 2014 in the Institute of Inorganic Chemistry at the University of Regensburg, under the supervision of Prof. Dr. Manfred Scheer.

Parts of this work have already been published:

- S. Heinl, S. Reisinger, C. Schwarzmaier, M. Scheer: Selective Functionalization of $P_4$ by Metal-Mediated C-P Bond Formation; *Angew. chem.* **2014**, 126, 7769; *Angew. chem. Int. Ed.* **2014**, 53, 7639.
- S. Reisinger, M. Bodensteiner, E. Moreno Pineda, J. McDouall, M. Scheer, R. A. Layfield: Addition of pnictogen atoms to chromium(II): synthesis, structure and magnetic properties of a chromium(IV) phosphide and a chromium(III) arsenide; *Chem. Sci.* **2014**, 5, 2443-2448.
- S. Scheuermayer, F. Tuna, M. Bodensteiner, M. Scheer, R. A. Layfield: Spin crossover in phosphorus- and arsenic-bridged cyclopentadienyl-manganese(II) dimers; *Chem. Commun.* **2012**, 48, 8087-8089.
- S. Scheuermayer, F. Tuna, E. Moreno Pineda, M. Bodensteiner, M. Scheer, R. A. Layfield: Transmetalation of Chromocene by Lithium-Amide, -Phosphide, and –Arsenide Nucleophiles; *Inor. Chem.* **2013**, 52, 3878-3883.

To Bernd

"A day without laughter is a day wasted."

– Charlie Chaplin

# Table of Contents

# Preface

"Cold fire", "Ignis perpetuus", "Phosphorus mirabilis": these are only a few of the names given to the luminous substance the Hamburg alchemist Hennig Brand had stumbled upon when distilling urine in 1669.[1] They illustrate very well how fascinated he and his contemporaries were, by what is known to us today as "bearer of light"–phosphorus. Still, the fifteenth element of the Periodic System of Elements has not lost its attraction, merely due to its reaction properties, that are beneficial for a very varied chemistry. This thesis, therefore, contains three research topics with each of them having a rather different chemical background, which is why, after the following general remarks on the heavier pnictogens phosphorus and arsenic, each chapter will start with a more specific introduction.

Elemental phosphorus exists in three different allotropes which are named after their colour white, red and black (*cf.* Scheme A). The first modification found was white phosphorus (**A**), a waxy white substance that is light- and air-sensitive and consists of molecular $P_4$ tetrahedra.[2] The character of red phosphorus has not been fully clarified yet, because it seems to be a mixture of different polymeric structures with varying long range order. It probably can be divided into five polymorphs (I-V) that can be derived from the commercially available, amorphous type I.[3] Due to the problem of single crystal growth for X-ray diffraction analysis, little is known about the structures of type II[4] and III. Fibrous red phosphorus (type IV)[5] and violet or Hittorf´s phosphorus (type V, **B**)[6] both contain tubes of alternating $P_8$ and $P_9$ cages which are connected by $P_2$ units. But they differ in the way two of the tubes are linked with each other via their $P_9$ unit: The fibres in type IV are formed because of the parallel connection whereas the crosswise arrangement in type V leads to double layers. Black phosphorus (**C**, **D**) is thermodynamically the most stable modification at ambient conditions. Its structure is based on undulated layers of condensed six-membered rings in chair conformation.[7] More recent allotropes are two different nanorods reported by Pfitzner *et al.* which are built up on $P_{12}$ units.[8]

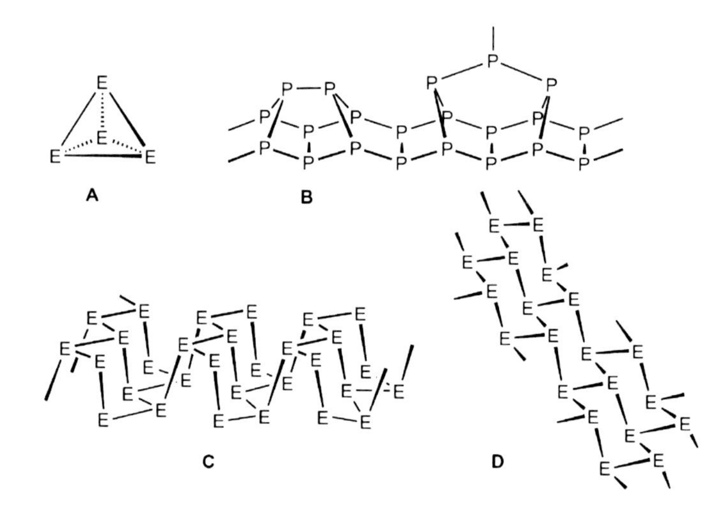


Scheme A: Structures[9] of A) white phosphorus and yellow arsenic, B) violet phosphorus, C) black phosphorus/arsenic, D) rhombohedral (high-pressure) black phosphorus/ grey arsenic (E = P, As)

Phosphorus´ heavier homologue arsenic is said to be discovered by Albertus Magnus around 1250.[10] Yellow arsenic (A) resembles white phosphorus as it consists of tetrahedral $As_4$ molecules, but it is much more sensitive to elevated temperatures, air and light[11] which makes it difficult to synthesise and almost impossible to store.[10] And that is why the several modifications[12] of yellow arsenic found between Bettendorf´s first description[13] and today are still not well investigated. The amorphous black modification can be obtained when arsenic vapour is condensed on warm (100-200 °C) surfaces.[14] It will transform into the metastable orthorhombic black arsenic (C)[15] when heated in mercury and easily forms mixed crystals with its analogue, black phosphorus.[16] Another but more dense allotrope consisting of condensed $As_6$ rings which are puckered to form undulated double layers, is grey arsenic (D).[17] The distance between As atoms of neighbouring layers is almost as short as within the layer, and each As atom is coordinated by six others in a distorted octahedral environment. In this way the structure resembles the cubic packing of metals. This, also called metallic arsenic, is the most stable modification at room temperature.

Even though both arsenic and phosphorus are essential elements,[9] and the latter makes with about 850 g a surprising share of the human body,[18] they are not that readily available for chemical reactions as one might think. Red phosphorus and grey arsenic are thermodynamically quite stable and very harsh conditions are needed for their activation. In academic research, for example, these two allotropes are reacted with sodium-potassium-alloy and trimethylsilylchloride (TMSCl) at high temperatures to give $E(SiMe_3)_3$ (E = P, As) which is a good starting material for reactions under mild conditions. Yellow arsenic is more reactive than the grey modification but the difficult access and the great sensitivity prevent it from industrial application. Therefore, metallic arsenic is used to get the demanded starting materials $As_2O_3$, $AsH_3$ or $AsCl_3$. In contrast, the phosphorus source of choice is, of course, white Phosphorus, which is very well illustrated by the large quantity of 500,000 tons that is produced per year. Its high reactivity notwithstanding, to get the industrial target products like detergents and other organophosphorus compounds, the $P_4$ tetrahedron has to be activated with aggressive methods like alkali metals, Grignard reagents or chlorine gas. To spare the dangerous and unsustainable detour via $PCl_3$ or $AsCl_3$ and their derivatives, there is a great interest in finding milder ways to provide the heavier elements of the group 15 for chemical reactions. A successful approach is the activation with transition metal[19] and main group[20] element fragments. Especially the former show promising properties regarding the development of catalytic species for the needed P–H and P–C bond formation directly from $P_4$. Many steps forward have been made, but still basic research has to be done to understand the mechanisms of phosphor degradation and transformation within the coordination sphere of a transition metal.

Due to that great interest, over the last decades phosphorus chemistry has become a large and versatile research field. Hence the following three chapters deal with rather different aspects of the organometallic chemistry of phosphorus and arsenic. First, contributions to the supramolecular chemistry of $P_n$ ligand complexes that are herein based on [Cp*Fe($\eta^5$-$P_5$)] and [Cp*Fe($\eta^5$-${}^{i}Pr_3C_3P_2$)] are shown, followed by an iron mediated activation of $P_4$ that results in a C-P bond formation in the second section, while the third chapter treats the use of phosphorus and arsenic as donor atoms in complexes with paramagnetic metal ions.

# 1. Coordination chemistry of $P_n$ ligand complexes

The story of transition metal complexes with unsubstituted, so called naked, phosphorus ligands, started in the seventies of the last century. The pioneers Ginsberg and Lindsell, who reported the first $P_n$ ligand complex in 1971,[21] were soon followed by the work of Markò[22] and Sacconi[23] (Scheme B). From then on, a plethora of transition metal compounds bearing $P_n$ ligands with n = 1-24 has been synthesised.[19-20, 24]

L = $P(m\text{-Tol})_3$, $P(p\text{-Tol})_3$, $PPh_3$, $AsPh_3$

M = Ni, Pd

Ginsberg and Lindsell 1971 — Markó *et al.* 1973 — Sacconi *et al.* 1979

Scheme B: First $P_n$ ligand complexes.

However, not only the realisation of different combinations of transition metal, size and coordination mode of the phosphorus moieties has been of interest, but, already at the beginning, the reactivity of the phosphorus lone pair has been investigated.[25] The Lewis acidic group six pentacarbonyl fragments are often the first choice: for example, the reaction of $[(Cp^*{}_2W_2(CO)_4(\mu,\eta^{2:2}\text{-}P_2)]$ with $[W(CO)_5{\cdot}thf]$ lead to $[(Cp^*{}_2W_2(CO)_4(\mu,\eta^{2:2}\text{-}P)(\mu,\eta^{2:2:1}\text{-}P)W(CO)_5]$[26]–the mono-coordinated product. And though with $[Mo(CO)_4(nbd)]$ (nbd = norbornadiene, $C_7H_8$), it was possible to connect $[(Cp_2Cr_2(CO)_4(\mu,\eta^{2:2}\text{-}P_2)]$ and $[(CpCr)_2(\mu,\eta^5\text{-}P_5]$ (Scheme C).[27] First attempts to use $P_n$ ligand complexes as building blocks in the supramolecular chemistry have been made by Midollini, Stoppioni and Peruzzini. They succeeded in the connection of two $[(triphos)M(\eta^3\text{-}P_3)]$ (triphos = 1,1,1-tris(diphenylphosphinomethyl)ethane; M = Co, Rh, Ir) units with mono-valent copper, silver and gold (Scheme C).[28]

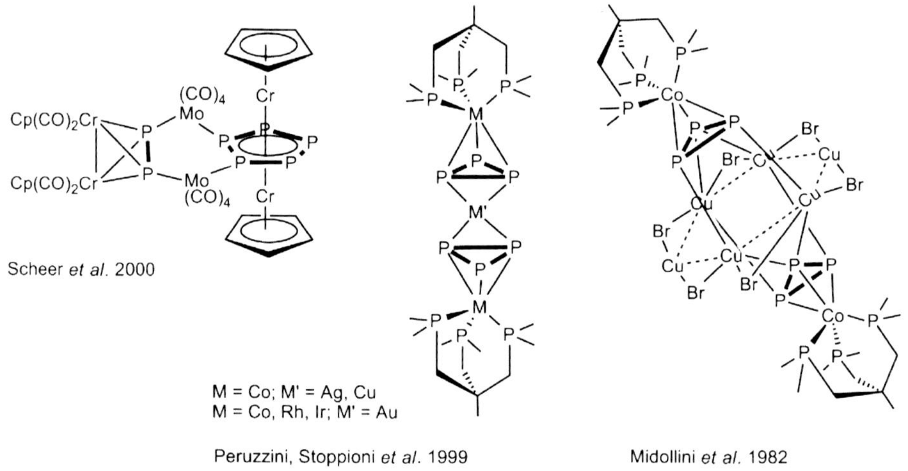

Scheer *et al.* 2000

M = Co; M' = Ag, Cu
M = Co, Rh, Ir; M' = Au

Peruzzini, Stoppioni *et al.* 1999

Midollini *et al.* 1982

**Scheme C:** Selected examples of supramolecular aggregates of $P_n$ ligand complexes and Lewis acidic fragments.

## 1.1 Spherical supramolecules from $[Cp^*Fe(\eta^5\text{-}P_5)]$ (1) and copper(I) halides

The introduction of coinage metal ions into this field allowed the realisation of larger aggregates–besides dimers, our group has realised from tetramers up to polymers built of complexes containing polyphosphorus ligands with two to six atoms.[25] Admittedly, the firstly by Scherer synthesised pentaphosphaferrocene[29] plays a special role: the cyclic arrangement of the phosphorus atoms in the *cyclo*-$P_5$ end-deck makes on the one hand a coordination with subsequent connection to further $[Cp^RFe(\eta^5\text{-}P_5)]$ molecules in all directions possible, and on the other hand already bears the five-membered ring motif enabling the composition of spherical aggregates (see Scheme D). Therefore, besides one- and two-dimensional polymers, the formation of fullerene-like 'nanoballs' made of $[Cp^*Fe(\eta^5\text{-}P_5)]$ (**1**) and copper(I) halides (Cl, Br, I) was achieved. The first supramolecule of this kind–$[Cp^*Fe(\eta^5\text{-}P_5)]@[\{Cp^*Fe(\eta^5\text{-}P_5)\}_{12}(CuX)_{10}(Cu_2X_3)_5\{Cu(CH_3CN)_2\}_5]$ (X = Cl, Br)–was reported 2003 and consists of two $C_{60}$ analogue half-shells linked by a belt of $Cu_2X_3$ and $Cu(CH_3CN)$ units, 90 non-carbon core atoms altogether.[30] The interior space of this spherical framework is not empty, but occupied by one molecule of **1**. To find out whether a template with a five-fold symmetry is needed for the building of such an aggregate, $C_{60}$ was offered to the system. The exclusive result was a spherical molecule with 99 inorganic core atoms incorporating one buckminsterfullerene: $C_{60}@[\{Cp^*Fe(\eta^5\text{-}P_5)\}_{13}Cu_{26}Cl_{26}(H_2O)_2(CH_3CN)_9]$.[31] Moreover, the reaction of $[Cp^*Fe(\eta^5\text{-}P_5)]$ (**1**) and CuCl in the presence of *ortho*-carborane ($o$-$C_2B_{10}H_{12}$) lead to its encapsulation in an icosahedral cavity with 80 core atoms ('80 vertex ball') that resembles the $I_h$-$C_{80}$ fullerene.[32] To further investigate the relation between the symmetry of the guest and the host molecule, templates with different symmetries were tested. Thus, white phosphorus and yellow arsenic–the tetrahedral molecular modifications of these elements–were added to the reaction mixture of **1** and CuI and were successfully enclosed in a supramolecule. The building of $E_4@[\{Cp^*Fe(\eta^5\text{-}P_5)\}_{10}(CuI)_{30}(CH_3CN)_6]$ (E = P, As) has not only been unexpected because of the halide–copper(I) iodide had never before assembled spherical structures–but also because of its cuboid shape.[33] A symmetry-fitting tetrahedral capsule was achieved for arsenic in $\{Z_4As_4\}@[\{Cp^*Fe(\eta^5\text{-}P_5)\}_{12}Cu_{51}I_{56}(CH_3CN)_3]^-Z^+$ (Z = light atom), however. The nature of the light atom Z could not yet be determined with absolute certainty, but an excess of LiCl in the reaction mixture, the charge balance, and the results of $^7$Li NMR measurements suggest that the atom in question is probably lithium.[34]

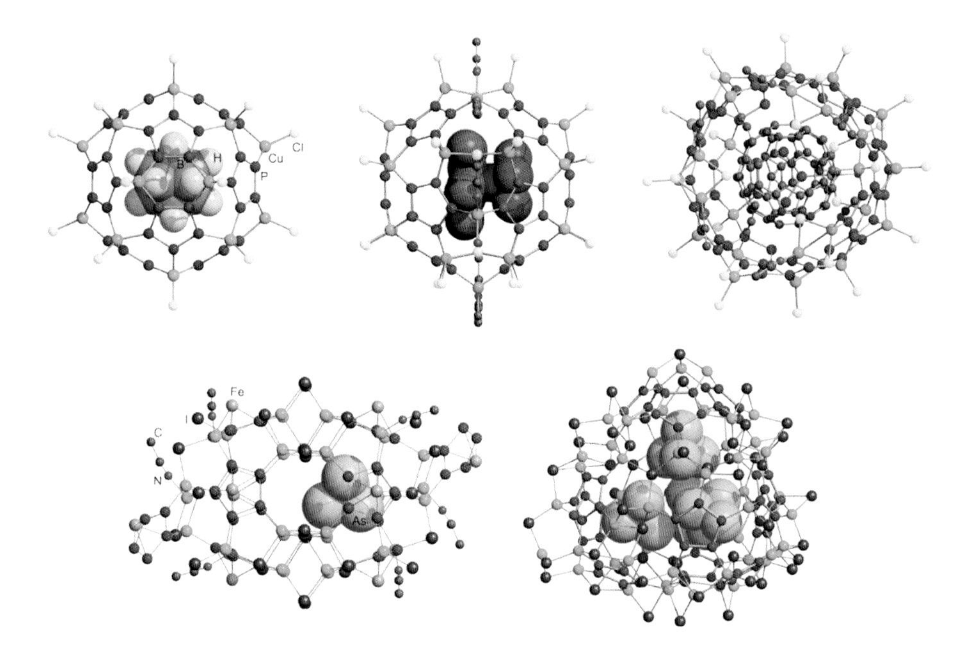


**Scheme D:** Selected examples of spherical aggregates consisting of $[Cp^*FeP_5]$ and copper(I) halides.

All of these observations suggest that the organisation of **1**, copper(I) halides and acetonitrile to ball-like structures strongly depends on the interactions with the guest molecule and therefore with its symmetry. Hence, the question whether such interactions are inalienable for the self-assembly of the inorganic cage, and the enclosed compound is a template in the truest sense of the word, is intriguing. To find an answer, adamantane ($C_{10}H_{16}$) was chosen as a guest molecule.

This cycloalkane with approximate tetrahedral symmetry seemed not much likely to interact strongly either with the $P_5$ rings or the copper(I) halide units. But it is small enough (∅: 7.4 Å[35]) to fit into the already known spherical frameworks (inner diameter: 80 inorganic core atoms–8.2 Å, $90_{CuCl}$ i. c. a.–12.5 Å, $90_{CuBr}$ i. c. a.–13.2 Å, 99 i. c. a.–13.5 Å).[36]

The reaction of [Cp*Fe($\eta^5$-$P_5$)] (**1**) with two equivalents of CuX (X = Cl, Br, I) in the presence of adamantane affords under a layering procedure single crystals of $C_{10}H_{16}$@[{Cp*Fe($\eta^5$-$P_5$)}$_{12}$(CuCl)$_{20-y}$] (**2**) and $C_{10}H_{16}$@[{Cp*Fe($\eta^5$-$P_5$)}$_{12}$(CuBr)$_{20-y}$] (**3**), respectively, in yields of 68 % (**2**) and 50 %(**3**) (Equation 1). The index y represents the vacancies in the scaffold (see below), elemental analysis (C, H- value) suggests y = 3, thus one supramolecule contains 17 CuX units on the average. CuI does not lead to a spherical coordination compound, but only to the already known $^2_\infty$[{Cp*Fe($\eta^5$-$P_5$)}(CuI)] ($P_{2d}$).[37]

**1** + 2 CuX → (adamantane) $C_{10}H_{16}$@[{Cp*Fe($\eta^5$-$P_5$)}$_{12}$(CuX)$_{20-y}$] (1)

**2**: X = Cl, 68 %
**3**: X = Br, 50 %

$C_{10}H_{16}$@[{Cp*Fe($\eta^5$-$P_5$)}$_{12}$(CuCl)$_{20-y}$] (**2**) crystallises as dark brown blocks from $CH_2Cl_2/CH_3CN$ in the trigonal space group $R\bar{3}$, from toluene/$CH_3CN$ in the triclinic space group $P\bar{1}$. The asymmetric unit contains one half and one sixth molecule or three half molecules, respectively. $C_{10}H_{16}$@[{Cp*Fe($\eta^5$-$P_5$)}$_{12}$(CuBr)$_{20-y}$] (**3**) crystallises with the same habitus from toluene/$CH_3CN$ in the triclinic space group $P\bar{1}$. Three half molecules are the crystallographically independent part.

The structure refinement of **2** and **3** has not been successfully finished up to the present. As already indicated by the index y in the chemical formula, the supramolecules are not complete. The crystal is merely the solid solution of spherical molecules with a varying amount of vacancies and therefore varying composition. Solvent molecules fill the space between the polynuclear complexes as well as the vacancies generated by the missing copper(I) halides, and act as an anchor to fix the balls in their orientation. This makes the obtained data very difficult to interpret. Nevertheless, the X-ray structure analysis reveals the heavy atom framework of **2** and **3** and shows that the compounds are based on the same connectivity pattern of CuX and [Cp*Fe($\eta_5$-$P_5$)]. Therefore, only **3** is depicted in Figure 1.

The general assembly of twelve pentaphosphaferrocene **1** that are coordinated through all five phosphorus atoms by 20-y CuX (X = Cl, Br) units is analogous to the known, complete icosahedral cluster with 80 non-carbon core atoms ('80 vertex ball') firstly observed encapsulating *ortho*-carborane (*cf.* Scheme D).[32] Therein, one $Cu^{I}$ ion connects three **1** molecules, a total of 30 $\{(P_2)_2Cu_2\}$ six-membered rings are formed between the $\{P_5\}$ moieties, so that the scaffold follows the isolated pentagon rule and its constitution therefore resembles the $C_{80}$ fullerene. The supramolecules have an outer diameter of approximately 21.7 Å (**2**) and 21.2 Å (**3**), respectively. The inner cavity has a diameter of approximately 7.7 Å in both **2** and **3**,[35] and the void is occupied by one adamantane molecule. Unfortunately, its refinement has not been finished successfully so far, due to a very high disordering. Apparently, the cycloalkane can reorientate in its cage without any predominant orientation.

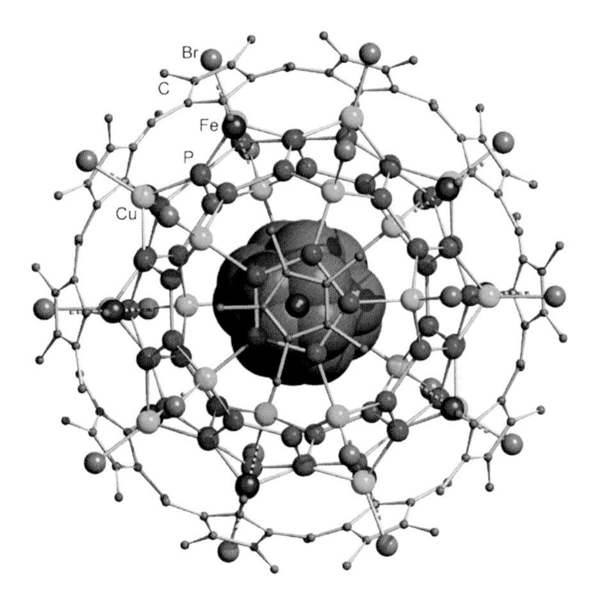


**Figure 1:** One of the molecules in the solid solution of **3** in the crystal. H atoms are omitted and C atoms represented with reduced radii for clarity reasons. Selected bond lengths [Å]: P(61)–P(62) 2.095(5), P(61)–P(65) 2.110(4), P(63)–Cu(18) 2.291(4), P(62)–Cu(17) 2.301(3), Cu(20)–Br(20) 2.322(2), Cu(11)–Br(11) 2.338(2).

Because the X-ray data do not unambiguously proof the nature of the molecule enclosed, additional characterisation methods were needed. Analysis via mass spectrometry was hampered because the molecular formulae of $Cp^{*\cdot}$ and adamantane only differ by one hydrogen atom that could easily be lost or absorbed ($Cp^{*\cdot}$ = $C_{10}H_{15}$ = 135.23 g/mol, adamantane = $C_{10}H_{16}$ = 136.23 g/mol). A direct detection of $C_{10}H_{16}$ inside the ball in solution with NMR spectroscopy is not possible either, due to the poor solubility of **2** and **3**. Thus, pyridine-$d_5$ was chosen as a solvent that breaks the coordination scaffold and thereby dissolves all components of the

supramolecules. Consequently, the $^{1}H$ NMR spectrum of **2** (and **3**) reveals a singlet at $\delta$ = 1.34 ppm ($\delta$ = 1.36 ppm) for the $CH_3$ groups of the Cp* ligands and two singlets at $\delta$ = 1.66 and 1.83 ppm ($\delta$ = 1.65 and 1.81 ppm) for the $CH_2$ and CH protons of adamantane. Furthermore, solid state MAS-NMR spectroscopy provided a possibility to detect the guest molecule inside its intact cage. The $^{1}H$ MAS NMR spectrum of **3** shows a very broad signal for the $CH_3$-groups of the Cp* ligands at $\delta$ = 2.1 ppm as well as a singlet for adamantane at $\delta$ = -3.1 ppm. The measurement of **2** reveals a very broad signal at $\delta$ = 1.7 ppm for the methyl groups that superimposes the adamantane signal, which only appears as a shoulder at $\delta$ = -3.2 ppm. Compared to the pure cycloalkane, which also shows only one signal in the $^{1}H$ MAS NMR spectrum at $\delta$ = 1.8 ppm, the signal is upfield shifted in both compounds. This could either be explained by a transfer of electron density from the cage molecule to adamantane or by a shielding effect of the enclosing $[\{Cp^*Fe(\eta^5\text{-}P_5)\}_{12}(CuX)_{20}]$ scaffold. The same tendency has been found for the encapsulated *o*-carborane in $C_2B_{10}H_{12}@[\{Cp^*Fe(\eta^5\text{-}P_5)\}_{12}(CuCl)_{20}]$, in which a charge transfer occurs.[32]
It is undoubted that examples like the icosahedral ball $C_2B_{10}H_{12}@[\{Cp^*Fe(\eta^5\text{-}P_5)\}_{12}(CuCl)_{20}]$[32] incorporating *ortho*-carborane or the tetrahedral cage $\{Z_4As_4\}@[\{Cp^*Fe(\eta^5\text{-}P_5)\}_{12}Cu_{51}I_{56}(CH_3CN)_3]^- Z^+$ (Z = light atom) enclosing $As_4$[34] suggest that the interactions between the building units of the host complex and the guest molecule have influenced the formation and the character of these compounds. These reactions are therefore prime examples of template controlled processes.[38]
But the successful trapping of adamantane in **2** and **3** inside the supramolecule with 80 vertices which has already acted as container for other template molecules as well (*o*-$C_2B_{10}H_{12}$, $[CpV(C_7H_7)]$, $P_4S_3$[39]), together with the lack of predominant orientations of the guest molecule inside the cavity, indicates that probably very weak interactions are sufficient to promote the assembly of such spherical aggregates. Consequently, at least for an encapsulation into this very host molecule, the size and the solubility are the most important properties for the selection of future guest molecules.

The system of pentaphosphaferrocene (**1**) and copper(I) halides is quite robust regarding the coordination products in the solid state which are mostly air- and moisture stable over a long period of time. But a high sensitivity can be observed regarding the reaction conditions. Small changes of the solvents, the concentration of the starting material, reaction time or temperature can make a big difference: the diffusion experiment of one equivalent of each **1** and $P_4$ in toluene with one equivalent CuCl is usually carried out at room temperature and leads to the one-dimensional polymer ${}^{1}_{\infty}[Cp^*Fe(\eta^5\text{-}P_5)CuCl]\cdot P_4$.[33] In a hot summer, however, the same reaction product can only be gained when the reaction is carried out in the refrigerator (6 °C).

To obtain crystals of good quality of the often insoluble coordination compounds, a slow crystal growth is necessary and achieved by a gradual reaction of the two components. This is ensured by carefully layering the solutions of the starting materials in a thin schlenk tube, where they slowly react with each other as the solutions merge. During this process the concentrations of all components–reactants as well as solvents–vary strongly especially within the boundary layer where the reaction takes place. Therefore, it is not very surprising that in the starting phase of a layering experiment one can often observe compounds in the form of powders for example that disappear again when the conditions changed enough with the state of mixing.

One of these compounds occurs during the reaction of $[Cp^*Fe(\eta^5\text{-}P_5)]$ (**1**) and CuI in ortho-dichlorobenzene and acetonitrile, finally leading to the formation of $^1_\infty[\{Cp^*Fe(\eta^5\text{-}P_5)\}\{CuI\}]\cdot 0.5C_6H_4Cl_2$ ($\mathbf{P_{odCB}}$) and $\mathbf{P_{2d}}$.[37, 40] When the two solutions have merged about one third, a few crystals of $(C_6H_4Cl_2)@[\{(Cp^*Fe(\eta^5\text{-}P_5)_8Cu_{24}(I)_4(\mu\text{-}I)_{14}(\mu_3\text{-}I)(\mu_4\text{-}I)_2(CH_3CN)_9\{Cu_4(\mu\text{-}I)_3(\mu_3\text{-}I)_3(\mu_4\text{-}I)(CH_3CN)\}]$ (**4**) can be found at the wall of the schlenk tube in the boundary layer region. They have vanished again when two thirds of the solutions are mixed and the final reaction products $\mathbf{P_{odCB}}$ and $\mathbf{P_{2d}}$ appear for the first time (Equation 2). This hampered the characterisation of **4** with further methods.

$$\mathbf{1} + CuI \xrightarrow[\text{r.t.}]{o\text{-}C_6H_4Cl_2/CH_3CN} \underset{\mathbf{4}}{C_6H_4Cl_2@[\{Cp^*Fe(\eta^5\text{-}P_5)\}_8(CuI)_{28}(CH_3CN)_{10}]} + \mathbf{P_{2d}} + \mathbf{P_{odCB}} \quad (2)$$

Compound **4** crystallises as red platelets in the triclinic space group $P\bar{1}$ with one molecule as the crystallographically independent part. Admittedly, the measurement is incomplete because of technical problems. A complete data set could unfortunately not be recorded within the scope of this thesis.

Nevertheless, X-ray structure analysis reveals a spherical supramolecule that consists of eight pentaphosphaferrocene (**1**) units, 24 copper and 21 iodine atoms and to which an additional $Cu_4I_7$ cage is connected. The $\{P_5\}$ moieties build the corners of a square antiprism with a side length of about 2.9 Å[35] which bears $Cu_4I_5$ caps at the top and at the bottom (Figure 2).

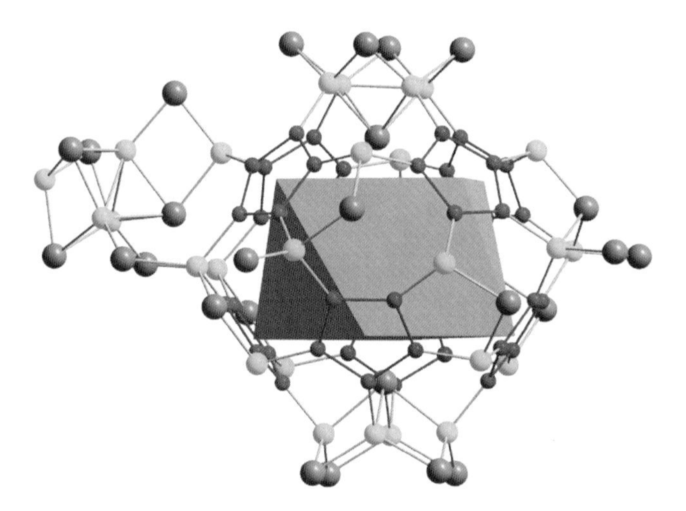

**Figure 2:** Representation of the antiprismatic shaped cavity of **4**.

In the middle, three [Cp*Fe($\eta^5$-$P_5$)] units are linked by three copper atoms to form a $Cu_3P_6$ nine-membered ring. In this, two Cu atoms are bridged by one iodide ligand and saturated either by another terminal iodide or an acetonitrile ligand. One of these $Cu_3P_6$ units is canopied by a {$\mu_3$-I} ion. The opposite $Cu_3P_6$ ring is linked via three iodide ligands instead to an additional copper-iodide cage. Its four copper atoms are arranged as a tetrahedron whose triangular faces are capped by iodine ligands and the apical Cu bears an acetonitrile ligand. The inner cavity has dimensions of 8.0 Å (distance between the inner I atoms of the $Cu_4I_5$ caps) by 12.3 Å long (distance between the opposite equatorial Cu atoms) and is occupied by one *ortho*-dichlorobenzene molecule (Figure 3).[35]

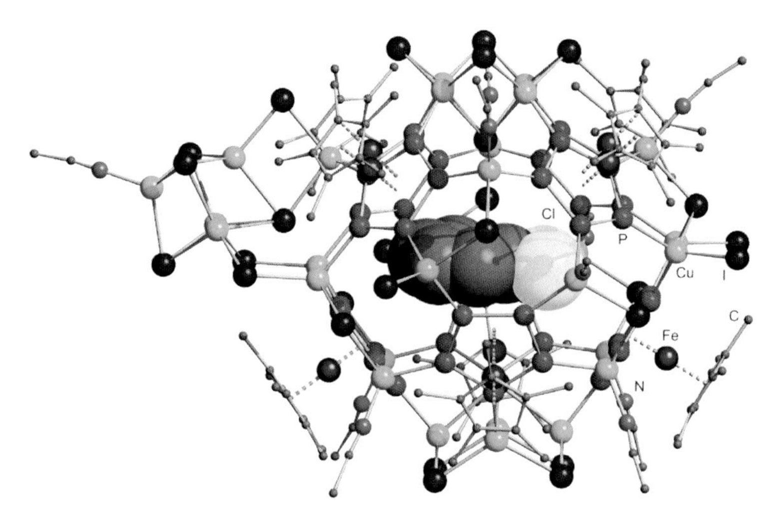


Figure 3: Molecular structure of **4** in the crystal. For clarity reasons, H atoms are omitted and the carbon atoms of the Cp* and acetonitrile ligands are represented with reduced radii. Selected bond lengths [Å]: P(53)–P(54) 2.0780(1), P(53)–P(52) 2.1195(1), Cu(17)–P(51) 2.2701(1), Cu(19)–P(53) 2.3478(1), Cu(16)–I(17) 2.5801(1), Cu(17)–I(16) 2.6948(1).

The inorganic framework of **4** is similar to that of $[\{Cp^{Bn}Fe(\eta^5\text{-}P_5)\}_8Cu_{24}(\mu\text{-}I)_8(\mu_3\text{-}I)_8(\mu_4\text{-}I)_2]$ ($Cp^{Bn}$ = penta-benzylcyclopentadienyl) synthesised in our group by F. Dielmann. There, all $Cu_3P_6$ motifs are capped by a $\{\mu_3\text{-}I\}$ unit and therefore divided into three $P_2Cu_2I$ five-membered rings. Furthermore, it does not bear an additional copper-iodide cage and co-crystallises with $[\{Cp^{Bn}Fe(\eta^5\text{-}P_5)\}_{12}\{CuI\}_{20}]$.[24]

Similar scaffolds with copper(I)-chloride or bromide are not known. But this is not that surprising as CuI is sometimes regarded as the 'black sheep' of the $Cu^I$ halides, because it tends to the formation of structural motifs that differ from the coordination compounds observed with CuCl or CuBr and pentaphosphaferrocene. Even though for example the two-dimensional polymer $^2_\infty[\{Cp^*Fe(\eta^5\text{-}P_5)CuX]$[37] as well as the spherical complex $[\{Cp^{Bn}Fe(\eta^5\text{-}P_5)\}_{12}(CuX)_{20}]$[24] (X = Cl, Br, I) have been realised with all three copper(I)-halides, there is still a group of assemblies whose members–like $E_4@[\{Cp^*Fe(\eta^5\text{-}P_5)\}_{10}Cu_{30}I_{30}(CH_3CN)_6]$ (E = P, As),[33] $^1_\infty[\{Cp^*Fe(\eta^5\text{-}P_5)\}\{CuI\}]\cdot 0.5C_6H_4Cl_2$,[40] or $[\{Cp^{Bn}Fe(\eta^5\text{-}P_5)\}_{12}Cu_aI_b(CH_3CN)_c]$[24] (a = 36-62, b = 32-56, c = 6-12), just to name a few–are solely known with iodide. Since the only structural relative is the above mentioned pentabenzyl derivative, the here presented copper-iodide ball belongs to this latter class.

## 1.2 Coordination polymers from [Cp*Fe($\eta^5$-$^i$Pr$_3$C$_3$P$_2$)] (5) and copper(I) halides

[Cp$^R$Fe($\eta^5$-P$_5$)] (R = organic substituent) is not only a potent and versatile building block for inorganic coordination polymers, but also a prime example of a P$_n$ ligand complex with all phosphorus atoms either bound to other phosphorus or to metal atoms. Compounds with organic ligands such as ferrocene, in which only some CR fragments are isolobally replaced by phosphorus do therefore, strictly speaking, not belong into this group. On the other hand, they are often accounted to it because of their very similar reactivity pattern towards the coordination at the phosphorus sites.[25]

This has well been demonstrated on phosphaferrocenes with a 1,2,4-triphospholyl-ligand. The groups of Nixon and Bartsch reported their coordination with various organometallic Lewis acids in the early 1990s, including the connection of two [Cp$^R$Fe($\eta^5$-$^t$Bu$_2$C$_2$P$_3$)] complexes by Re(CO)$_3$Br and Ni(CO)$_2$,[25, 41] respectively. The introduction of coinage metal salts by our group opened the door towards larger supramolecular assemblies and besides tri- and tetramers, one- and two-dimensional polymers have been realised (*cf.* Scheme E).[42]

**Scheme E:** Selected examples of supramolecular aggregates from triphosphaferrocene and copper(I)halides.

But surprisingly, the corresponding 1,3-diphosphaferrocene has not attracted that much attention yet. Even though, the di- and triphospholylsalts K[$^t$Bu$_3$C$_3$P$_2$] and K[$^t$Bu$_2$C$_2$P$_3$] are built together and cannot be separated when synthesised from phosphaalkene [(Me$_3$SiO)$^t$BuCP(SiMe$_3$)]

and K[P(SiMe$_3$)$_2$].[43] Our group investigated the isopropyl derivatives, K[$^i$Pr$_3$C$_3$P$_2$] and K[$^i$Pr$_2$C$_2$P$_3$], to develop metalphosphide precursers for the use in metalorganic chemical vapour deposition (MOCVD). In this way, the synthesis of these phospholyl salts had been established and the actual byproduct, [$^i$Pr$_2$C$_2$P$_3$]$^-$ became an object of interest for this thesis.

Reacting a mixture of these salts with a iron(II) halide (Cl, Br) and an alkalimetal cyclopentadienide gives [Cp$^R$Fe($\eta^5$-$R'_2$C$_2$P$_3$)] along with [Cp$^R$Fe($\eta^5$-$R'_3$C$_3$P$_2$)] (here Cp$^R$ = Cp*) (Equation 3).[41d, 43-44]

2 Me$_3$SiO(SiMe$_3$)C=P + K[P(SiMe$_3$)$_2$] —(toluene/Et$_2$O, r.t.)→ K[$^i$Pr$_3$C$_3$P$_2$] + K[$^i$Pr$_2$C$_2$P$_3$]

K[...] + K[...] + FeBr$_2$(dme) + LiCp* —(thf, -50 °C)→ ... **5** ... (3)

■ = C$^i$Pr

Copper(I) halides have so far proven to coordinate readily to a large variety of P$_n$ ligand complexes and to build up interesting frameworks by doing so. Thus, [Cp*Fe($\eta^5$-$^i$Pr$_3$C$_3$P$_2$)] (**5**) was reacted with two equivalents of CuX (X = Cl, Br, I) by a layering procedure in a 1:2-mixture of toluene and acetonitrile. In all three cases, a one dimensional polymer is formed (Equation 4): $^1_\infty$[(Cp*Fe($\eta^5$-$^i$Pr$_3$C$_3$P$_2$)Cu$_2$($\mu$-X)$_2$(CH$_3$CN)] (**6**: X = Cl, **7**: X = Br) and $^1_\infty$[(Cp*Fe($\eta^5$-$^i$Pr$_3$C$_3$P$_2$){Cu$_2$($\mu$-I)$_2$}(CH$_3$CN)$_{0.5}$] (**8**), respectively.

(4)

The chlorine derivative **6** could not be characterised by single crystal x-ray diffraction due to technical problems. But as the unit cell parameters (a = 15.744 Å, b = 19.294 Å, c = 20.384 Å, α = 90°, β = 91.7°,γ = 90°, V = 6168 $Å^3$), as well as the results of mass spectroscopic investigations and the elemental analysis correspond with those obtained for the bromine polymer **7** (a = 15.729 Å, b = 19.095 Å, c = 20.306 Å, α = 90°, β = 92.6°, γ = 90°, V = 6099 $Å^3$), **6** very probably isostructural to **7**.

Compound **7** crystallises as orange platelets in the monoclinic space group *C2/c*. The asymmetric unit contains $[\{Cp^*Fe(\eta^5\text{-}^iPr_3C_3P_2)\}Cu_2(\mu\text{-}Br)_2(CH_3CN)]$ and one additional acetonitrile molecule (Figure 4).

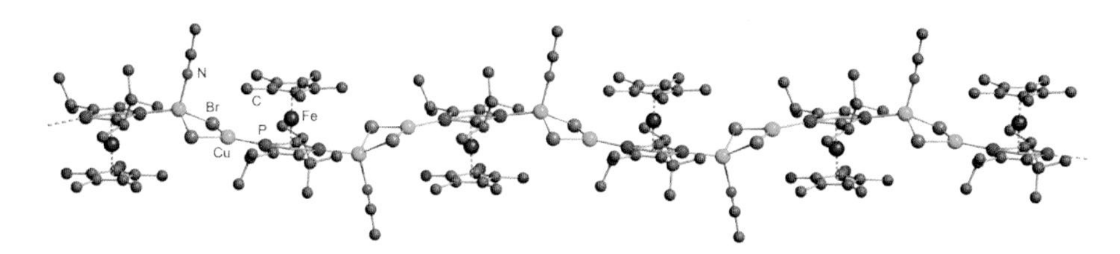


**Figure 4:** Section of the structure of **7** in the crystal. H atoms are omitted for clarity. Selected bond lengths [Å] and angles [°]: P(2)–C(1) 1.7516(1), P(2)–C(3) 1.7774(1), P(1)–Cu(1) 2.1733(4), P(2)–Cu(2) 2.2121(3), Cu(1)–Br(1) 2.3776(2), Cu(2)–Br(1) 2.4955(3), Cu(1)···Cu(2) 3.0475(2).

Compound **8** crystallises as well as orange platelets, but in the orthorhombic space group $P2_12_12_1$. [{Cp*Fe($\eta^5$-$^i$Pr$_3$C$_3$P$_2$)}$_2${Cu$_2$($\mu$-I)$_2$}$_2$(CH$_3$CN)] and two additional acetonitrile molecules are the crystallographically independent part (Figure 5).

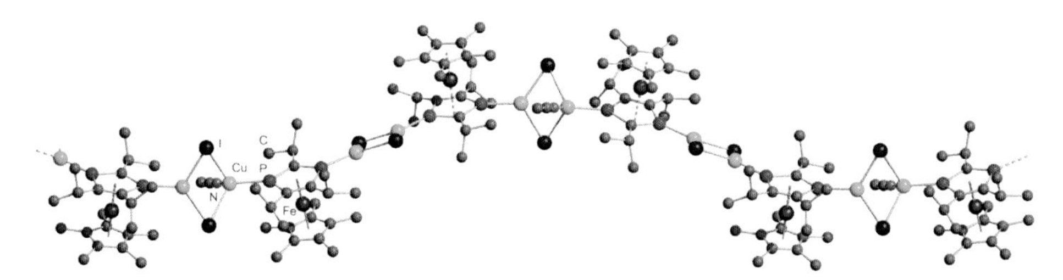


**Figure 5:** Section of the structure of **8** in the crystal. H atoms are omitted for clarity. Selected bond lengths [Å]: P(14)–C(75) 1.75(1), P(11)–C(38) 1.79(1), P(11)–Cu(5) 2.180(4), P(13)–Cu(7) 2.236(3), Cu(9)–I(2) 2.532(2), Cu(7)–I(1) 2.722(3), Cu(9)···Cu(7) 2.775(2), Cu(6)···Cu(5) 2.916(2).

X-ray structure analysis reveals that in both, **7** and **8**, the one-dimensional chains are built by connecting the [Cp*Fe($\eta^5$-$^i$Pr$_3$C$_3$P$_2$)] (**5**) molecules with $Cu_2X_2$ four-membered rings. The bending of the strands in **7** and **8** is caused by the position of the phosphorus atoms within the $P_2C_3$ five-membered ring and the tetrahedral coordination environment of some of the copper ions. The latter makes the difference between the two polymers: At first, they only differ in their amount of coordinated acetonitrile ligands. But this additional ligand leads to an expansion of the coordination environment of the copper centres from trigonal planar to tetrahedral. In the bromine derivative **7**, every second Cu atom is affected and thus, all {$Cu_2Br_2(CH_3CN)$} rings are similar. The iodide polymer, in contrast, contains less acetonitrile and only every fourth copper centre is coordinated in a distorted tetrahedral mode. Consequently, the connecting {$Cu_2I_2$} and {$Cu_2I_2(CH_3CN)$} units are perpendicular to each other. This results in a different undulation of the chains and the crystallisation in a chiral space group, whereas compound **7** crystallises in a centrosymmetric space group.

This coordination pattern of the copper(I) halides–$Cu_2X_2$ four-membered rings–which leads to the formation of one-dimensional polymers, is well known. It has been observed for many of supramolecular aggregates with $P_n$ ligand complexes (*cf.* Scheme F).[25]

**Scheme F:** Selected examples of one-dimensional polymers from $P_n$ ligand complexes and $Cu_2X_2$ units.

With only two opposite coordination sites, the scaffolds that can be built of diphospholyl ligands and Lewis acidic fragments are limited to a linear arrangement, unless other geometries are enabled by the contribution of the connecting units. Nevertheless, these polymers confirm that diphosphaferrocene (**5**) is as suitable for the coordination of copper(I) halides, which is accompanied by the self-assembling to larger supramolecular structures, as triphosphaferrocenes have shown to be beforehand (see above). This analogue reactivity pattern let it seem very likely that the phosphorus´ lone pairs in **5** are accessible as well for the coordination of other Lewis acids, which should be investigated in the future.

# 2. Iron mediated C-P bond formation starting from $P_4$

From flame retardants to herbicides, organophosphorus compounds are very important industrial products for our everyday life, and a large share of the annually produced white phosphorus has to undergo a hazardous halogenation, for a direct functionalisation with organic substituents is not possible yet. Therefore, the formation of P–C bonds from $P_4$, the first industrially manufactored product in this chain, is of great interest because of both, environment and safety reasons. However, though there is a veritable research history in this field, the desired catalytic activation process for white phosphorus remains a challenging task.

An early step were the reactions of $P_4$ with organomagnesium and -lithium compounds, which only lead to the non-selective formation of mixtures of the corresponding organophosphides.[45] Forty-five years later, Bertrand *et al.* reported the establishment of direct P–C bonds in good yields via the degradation,[46] aggregation[47] and fragmentation[48] of white phosphorus with stable N-heterocyclic carbenes. A different way to introduce a carbene as substituent was to react the frustrated Lewis acid/base pair $B(C_6F_5)_3$/N-heterocyclic carbene with $P_4$.[49] Here, only one P–P bond has been cloven to result in a tetraphospha-*bicyclo*[1.1.0]butane core (Scheme G).

Bertrand *et al.*

Dipp = $C_6H_3$-2,6-$^i Pr_2$

Tamm *et al.*

Scheme G: Examples of the activation of $P_4$ by stable carbenes.

While this initialising step of the degradation of the phosphorus tetrahedron is well known for transition metal fragments, only three more so-called butterfly molecules with direct P–C bonds have been synthesised in this way as far as we know. Fluck and co-workers reported in 1987 that the reaction of (sMes)Br and Li(sMes) (sMes = 2,4,6-$^{t}Bu_3$-($C_6H_2$)) with white phosphorus leads in minor yields to $sMes_2P_4$.[50] This molecule is the result of a nucleophilic attack of $R^-$ at the $P_4$ tetrahedron followed by the reaction with the electrophile (sMes)Br. The formation of [$\{Ar^{Dipp}\}_2P_4$] ($Ar^{Dipp}$ = $C_6H_3$-2,6-($C_6H_3$-2,6-$^{i}Pr_2)_2$) presented by Power *et al.* is achieved by the reaction of $Tl_2\{Ar^{Dipp}\}_2P_4$ with $I_2$.[51] The first radical activation was recently reported by Cummins *et al.*: $P_4$ consumes the Dmp radicals (Dmp = 2,6-$Mes_2C_6H_3$) generated in situ by the dehalogenation of DmpI with [Ti(N($^{t}$Bu)Ar)$_3$] (Ar = 3,5-$Me_2C_6H_3$) evolving $Dmp_2P_4$.[52] However, all organic substituents mentioned above have in common that the C–P bond is formed with $sp^2$ hybridised carbon atoms. Reactions involving $sp^3$ carbon reagents were lacking.

In our group, C. Schwarzmaier lately observed that [{Cp'''Fe(μ-Br)}$_2$] (**9a**) (Cp''' = $C_5H_2{}^{t}Bu_3$) reacts with white phosphorus at room temperature not only to [(Cp'''Fe)$_2$(μ,$\eta^{4:4}$-$P_4$)], but also to the organo-substituted butterfly compound [Cp'''$_2P_4$] (**10**) with $sp^3$-hybridised carbon atoms directly bonded to the phosphorus atoms.[34]

The following reaction mechanism (Equation 5) was proposed on the basis of our detailed observations of the reaction process and of Sitzmann's report on the generation of pentaisopropylcyclopentadienyl ($Cp^{iPr}$) radicals from Na($Cp^{iPr}$) and $FeCl_2$.[53] Although an explanation for the radical formation is missing, due to the fact that its derivatives **9a** and [{$Cp^{4iPr}$Fe(μ-Br)}$_2$] (**9b**) are observed in analogue reactions,[54] it is quite likely that [{$Cp^{iPr}$Fe(μ-Cl)}$_2$] occurs as an intermediate. This multistep mechanism suggests an $Fe^{III}$ species, namely [Cp'''$FeBr_2$], to be responsible for the formation of the $\{Cp'''\}^{\bullet}$ radicals and the butterfly compound, respectively. It has derived from the disproportionation of the $Fe^{II}$ starting compound **9a** in the presence of white phosphorus. The $P_4$ tetrahedron coordinates to the 16 VE iron(III) fragment and the bond to the Cp''' ligand is cleaved homolytically resulting in a $\{Cp'''\}^{\bullet}$ radical and the insoluble $FeBr_2$. The radical reacts with the coordinated $P_4$ and opens one of the P-P bonds to form **10**.

(5)

**9a** ■ = C*t*Bu

**10** 14 % (NMR)

Parallel, S. Heinl attempted to synthesise a $P_n$ ligand complex with copper following Scherer's preparation method.[55] But the reported in situ generation of [$Cp^{4iPr}$CuCO] that lead with $P_4$ to [$Cp^{4iPr}Cu(\eta^2-P_4)$] and [$Cp^{4iPr}Cu(\mu,\eta^{2:1}-P_4)CuCp^{4iPr}$], could not be transferred to the ligand $Cp^{BIG}$ ($Cp^{BIG} = C_5(4-{}^{nBu}C_6H_4)_5$). Already the addition of $NaCp^{BIG}$ to CuBr in THF (tetrahydrofurane, $C_4H_8O$) caused a dark blue colour that even persisted the discharging of CO gas into the solution. This strongly suggested the formation of $\{Cp^{BIG}\}^{\bullet}$ radicals, which was confirmed by EPR studies. However, these radicals reacted with white phosphorus to [$Cp^{BIG}{}_2P_4$] (**11**), a metal free butterfly-like compound with the $Cp^{BIG}$ ligands directly bonded to phosphorus atoms (Equation 6). Remarkable was further the $^{31}P$ NMR spectrum of **11**. It shows two coupled triplets at $\delta$ = -181.0 ppm and $\delta$ = -308.2 ppm as it is typical of butterfly $P_4$ ligands.[56] But at the same time, it resembles surprisingly well the $^{31}P$ NMR spectrum reported for the copper coordinated complexes [$Cp^{4iPr}Cu(\eta^2-P_4)$] and [$Cp^{4iPr}Cu(\mu,\eta^{2:1}-P_4)CuCp^{4iPr}$].[55]

2 $NaCp^{BIG}$ —(2 CuBr; - 2 $Cu_{metal}$, - 2 NaBr)→ 2 $\{Cp^{BIG}\}$ —($P_4$)→ $Cp^{BIG}$–$P_4$–$Cp^{BIG}$ (6)

**11** 29 %

These three intriguing observations, the iron mediated formation of [$Cp'''_2P_4$] (**10**) and the radical $P_4$ activation with $Cp^{BIG}$ to [$Cp^{BIG}{}_2P_4$] (**11**) as well as the inconsistent NMR spectra of Scherer's copper complexes, arouse a few questions: Is an iron(III) complex the active species leading to the formation of {Cp'''}$^{\bullet}$ radicals followed by the cleavage of one P-P bond to **10**? Can this reaction be transferred to other Cp ligands? And does the NMR signal reported for [$Cp^{4iPr}Cu(\eta^2-P_4)$] and [$Cp^{4iPr}Cu(\mu,\eta^{2:1}-P_4)CuCp^{4iPr}$] rather belong to the copper free compound [$Cp^{4iPr}{}_2P_4$] (**13**)? Answers to these questions should give the following investigations:

To generate [Cp'''$FeBr_2$] in situ, $FeBr_3$ is reacted with NaCp''' in toluene at room temperature. The addition of half an equivalent of white phosphorus leads to the formation of the desired [Cp'''$_2P_4$] (**10**) (Equation 7). The $^{31}P\{^1H\}$ NMR spectrum ($C_6D_6$) shows the exclusive formation of **10** with about 38% of conversion of $P_4$. Furthermore, the ratio of the four constitutional isomers can be found as reported for the iron(II) mediated reaction. The occurrence of only the four isomers with tertiary carbon atoms bound to the 'wingtip' phosphorus atoms resembles the higher stability of tertiary radicals, confirming the radical character of this reaction.

2 $FeBr_3$ + 2 $MCp^R$ —(toluene, r.t.)→ 2 [$Cp^RFeBr_2$] —($P_4$)→ $Cp^R{}_2P_4$ (7)

**10:** $Cp^R$ = Cp''', M = Na, 53 % (NMR)
**11:** $Cp^R$ = $Cp^{BIG}$, M = Na, 38 % (NMR)
**12:** $Cp^R$ = Cp*, M = Li, 82 % (NMR)
**13:** $Cp^R$ = $Cp^{4iPr}$, M = Na, 62 % (NMR)

The 'copper route' described above has been a well working way to {$Cp^{BIG}$}$^{\bullet}$ radicals and **11** as a consequence. Yet, it has not been possible to apply this method to other Cp ligands. It therefore seems that the $Fe^{III}$ fragment enables the formation and reaction of Cp radicals that are not stable as free radicals to the necessary extent. Furthermore, the yields have almost doubled–from 29% to 53%–by using $FeBr_3$ instead of CuBr. Still, the reaction with the un-substituted cyclopentadienyl ligand unfortunately cannot be realised, but the expansion to Cp* and $Cp^{4iPr}$ has been achieved.

In this way, the reaction of LiCp* and $FeBr_3$ with white phosphorus in toluene at ambient temperatures results in the formation of [$Cp^*{}_2P_4$] (**12**). The $^{31}P\{^1H\}$ NMR spectra ($C_6D_6$) show that 82% of the $P_4$ inserted has been converted to the Cp* substituted tetraphospha-*bicyclo*-butane which reveals two coupled triplets at $\delta$ = -144 ('wing-tip') and -367 ppm ('bridgehead') ($^1J_{PP}$ = 193 Hz). In the $^1H$ NMR spectrum ($C_6D_6$), three singlets are detected at $\delta$ = 1.02 (3H), 1.65 (6H) and 1.89 ppm (6H) for the methyl substituents at the Cp* ring.

Compound **12** is soluble in polar as well as non polar solvents and crystallises from a saturated $Et_2O$ solution as colourless blocks in the monoclinic space group $P2_1/c$ . X-ray structure analysis confirms the butterfly motif in a *syn,syn*-configuration. Compared to the other here discussed $[Cp^{R}_2P_4]$ compounds **10**, **11**, and **13**, however, the Cp* rings are twisted in the opposite direction, likely due to packing effects. The P–P and P–C bond distances lie in the expected range (*cf.* Figure 6).

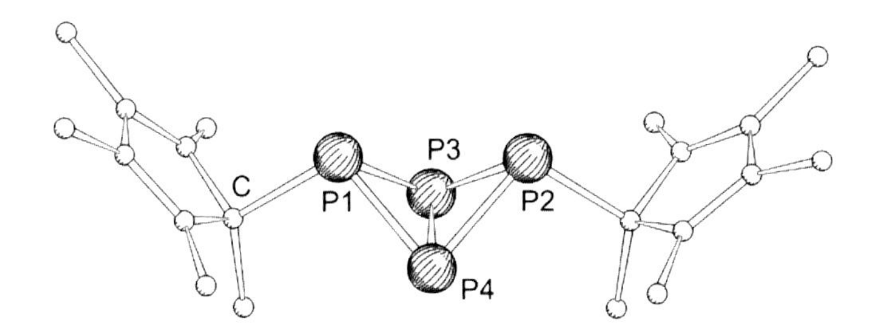


**Figure 6:** Molecular structure of **12** in the crystal. For clarity reasons, hydrogen atoms are omitted and carbon atoms diminished. Selected bond lengths [Å] and angles [°]: P1–P3 2.2177(4), P1–P4 2.2237(5), P2–P3 2.2131(4), P2–P4 2.2218(4), P3–P4 2.1924(4), P1–C 1.9073(14), P2–C 1.9175(14), P1···P2 2.8891(4), C–P1–P3 101.89(4), C–P1–P4 102.04(4), C–P2–P3 102.78(4), C–P2–P4 102.48(4), P1–P3–P2 81.39(2), P1–P4–P2 81.07(2), P3–P1–P4 59.16(1), P3–P2–P4 59.25(1);

$[Cp^{4iPr}{}_2P_4]$ (**13**) can be obtained from the analogue reaction of $NaCp^{4iPr}$ and $FeBr_3$ with $P_4$ in toluene. According to the $^{31}P\{^1H\}$ NMR spectrum ($C_6D_6$), 62% of the white phosphorus has been converted to **13**, and though several isomers are conceivable, only two can be observed. This might either be caused by steric hindrance or due to a superimposition of the signals. A $^{31}P\{^1H\}$-$^{31}P\{^1H\}$-COSY NMR experiment was performed to clarify the correlation of the two signal groups found between $\delta$ = -134 and -142 ppm as well as between $\delta$ = -311 and -366 ppm and the two isomers with an $A_2M_2$ and $A_2MN$ spin system, respectively (see Exp. section for details).

Crystals of **13** can be obtained from a saturated $Et_2O$ solution in form of colourless blocks. It crystallises in the triclinic space group $P\bar{1}$. X-ray structure analysis reveals the expected tetraphospha-*bicyclo*-butane scaffold in a *syn-syn*-configuration. The structural parameters lie within the range known for $Cp^{R}_2P_4$ compounds (*cf.* Figure 7).

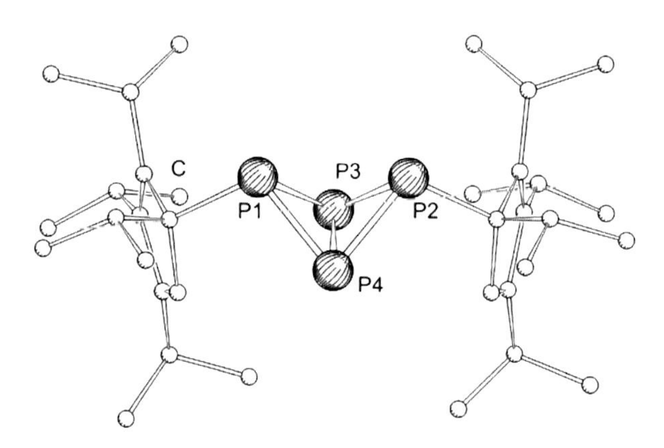


Figure 7: Molecular structure of **13** in the crystal. For clarity reasons, hydrogen atoms are omitted and carbon atoms diminished. Selected bond lengths [Å] and angles [°]:P1–P3 2.2124(8), P1–P4 2.214(5), P2–P3 2.2194(7), P2–P4 2.236(4), P3–P4 2.178(5), P1–C 1.889(3), P2–C 1.899(3), C–P1–P3 105.99(9), C–P1–P4 102.1(1), C–P2–P3 103.0(1), C–P2–P4 104.82(9), P1–P3–P2 77.6(1), P1–P4–P2 78.00(3), P3–P1–P4 58.5(1).

It is also noteworthy that the NMR data found for $[Cp^{4iPr}{}_2P_4]$ (**13**) resembles very much the signal pattern reported by Scherer for $[Cp^{4iPr}Cu(\eta^2\text{-}P_4)]$ and $[Cp^{4iPr}Cu(\mu,\eta^{2:1}\text{-}P_4)CuCp^{4iPr}]$.[55] Taken into account, that our synthesis is copper free and we have been able to characterise **13** further by FD-MS spectrometry (see exp. sec.) and X-ray structure analysis, there seem to be reasonable hints that Scherer´s group, unfortunately, misinterpreted their results and have therefore been the first to observe $[Cp^{4iPr}{}_2P_4]$ (**13**) unconsciously.

The butterfly compounds **10**, **11**, **12** and **13** were synthesised at ambient conditions from white phosphorus mediated by the in situ generated iron(III) fragment $[Cp^RFeBr_2]$ (equation 7). This method allows the selective activation of $P_4$ leading to a direct bond between phosphorus and a $sp^3$ hybridised carbon atom.

In addition to the fact that for **10** and **13** only isomers with tertiary carbon atoms bound to the 'wingtip' phosphorus are found, these observations fit the proposed radical mechanism[57] not only has it been possible to reproduce the already known butterfly molecules $[Cp'''_2P_4]$ (**10**) and $[Cp^{BIG}{}_2P_4]$ (**11**) but also to introduce Cp* and $Cp^{4iPr}$ to the system. This has finally helped to clarify the structures of $[Cp^{4iPr}Cu(\eta^2\text{-}P_4)]$ and $[Cp^{4iPr}Cu(\mu,\eta^{2:1}\text{-}P_4)CuCp^{4iPr}]$ or **13**, respectively. Following these examples, there is now a new way of making phosphorus available for organic synthesis. It is therefore surely worth further exploring possible substituents, that on the one hand can coordinate to an iron(III) centre and be turned into a radical there, and are on the other hand attractive targets for the synthesis of organic molecules. Furthermore, these new organo-substituted butterfly molecules can serve as $P_n$ ligands brought into the coordination sphere of an organometallic fragment. And as it comes to supramolecular chemistry, they might just as well become $P_n$ building blocks that can be linked by coinage metal salts to form interesting frameworks.

# 3. Magnetically active compounds with P- and As-based ligands

Electronic media play an important role in our society and have found their way into everyday life. Therefore, the devices have to suit the demands of flexibility and mobility by getting smaller and faster, whereas the data volume that needs to be handled constantly increases at the same time. Molecular magnetism can be a solution to that problem and caused a boost to the research on magnetically active coordination complexes. To design switchable materials on a molecular basis, it is necessary to gain more insight into the factors that influence the magnetic properties and how to control them.[58] One of these fascinating properties is the high-spin/low-spin bistability found in coordination complexes of the first-row transition metals. The spin states of those so called spin crossover (SCO) compounds can be switched by external stimuli like temperature, light, or pressure.[59]

Since the discovery of unusual magnetic behaviour of copper(II) acetate by Bleaney and Bowers in 1952,[60] there has been much research interest in multimetallic complexes and examples with a large variety of metal ions have been synthesised. However, the ligands, that bridge the metal centres to mediate the exchange coupling, mostly contain hard donor atoms like oxygen or nitrogen.[61] With the exception of thiolate ligands,[62] the influence of bridging soft donor atoms like the heavier pnictogens phosphorus and arsenic has not yet been investigated thoroughly. The relatively large and diffuse valence orbitals of heavier pnictogens could be beneficial for the overlap of metal and ligand orbitals, and hence could enhance the superexchange coupling. These considerations provide the starting point for investigating the potential of μ-phosphide and -arsenide ligands in coordination complexes with paramagnetic metal ions–a situation that had not previously been investigated.

In terms of open shell metal ions, manganese(II) aroused our interest because of its versatile coordination geometry and the high spin (HS) character.[9] Manganese(II) compounds with heavier pnictogens as bridging ligands are rare (Scheme T). In the few $Mn^{II}_2P_2$ systems studied so far, only the effective magnetic moments, $\mu_{eff}$, were determined.[63],[64],[65] Exchange coupling constants have hitherto only been studied quantitatively with quantum chemical methods by von Hänisch for $[Mn_5\{N(SiMe_3)_2\}\{\mu_4\text{-}PSi^iPr_3\}_2\{\mu\text{-}P(H)Si^iPr_3\}_5]$.[66] The lack of knowledge of exchange coupling in P-bridged $Mn^{II}$ complexes, and the fact that manganocenes are known for showing spin equilibria,[67] made this aspect an interesting material to investigate. Therefore, a phosphorus-bridged $Mn^{II}$ dimer with capping cyclopentadienyl ligands was chosen as target molecule.

The reaction of **14** and lithium-bis(trimethyl)silylphosphide and –arsenide, respectively, in toluene at ambient temperatures leads to the formation of the bimetallic complexes $[CpMn\{\mu\text{-}P(SiMe_3)_2\}]_2$ (**15**) and $[CpMn\{\mu\text{-}As(SiMe_3)_2\}]_2$ (**16**) (Equation 8). After filtering off the precipitated LiCp, crystals of **15** and **16**, respectively, can be obtained from saturated solutions at -28 °C.

$$2\ Cp_2Mn\ (\mathbf{14}) + 2\ LiE(SiMe_3)_2 \xrightarrow[\text{toluene}]{-78^\circ C \rightarrow r.t.} [CpMn\{\mu\text{-}E(SiMe_3)_2\}]_2 + 2\ LiCp \qquad (8)$$

**14**

**15** E = P, 49 %
**16** E = As, 37 %

The phosphorus derivative **15** crystallises as orange parallelepipeds in the triclinic space group $P\bar{1}$ with two molecules in the asymmetric unit (only one will be discussed in the following). Compound **16** crystallises with the same crystal habit in the monoclinic space group $P2_1/n$ (Figure 8).

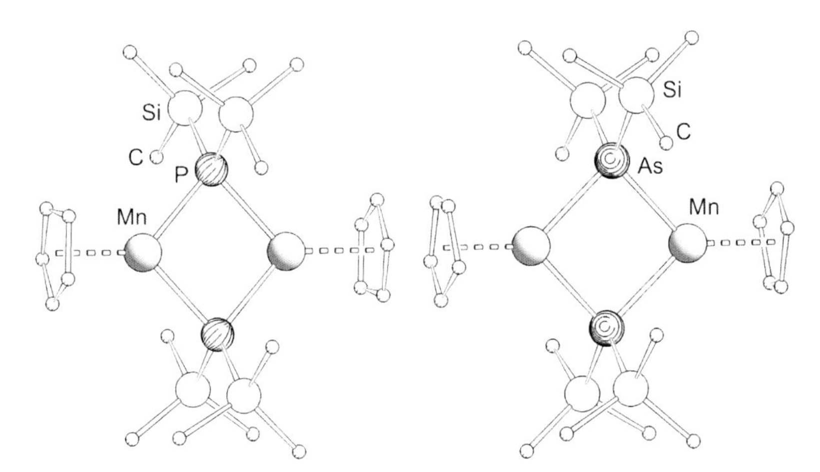


**Figure 8:** Molecular structure of **15** (left) and **16** (right) in the crystal. For clarity reasons, hydrogen atoms are omitted and carbon atoms are represented with reduced radii. Selected bond lengths [Å] and angles [°]: **15** a) Mn1–P1 2.3813(9), Mn1–P1A 2.3866(9), M···M 3.066, b) Mn1–P1 2.3858(9), Mn1–P1A 2.3866(9); P1–Mn1–P1A 99.97(3), Mn1–P1–Mn1A 80.03(3); **16** a) Mn1–As1 2.4963(5), Mn1–As2 2.4992(5), Mn1···Mn1A 3.394, As1–Mn1–As1A 94.41(2), Mn1–As1–Mn1A 85.59(1) b) Mn1–As1 2.5011(5), Mn1–As2 2.5010(4), Mn1···Mn1A 3.418, As1–Mn1–As1A 93.79(2), Mn1–As1–Mn1A 86.21(1).

The crystal structure analysis reveals that **15** and **16** each contain two manganese centres, that are formally five-coordinate by regarding the Cp ligand as occupying three coordination sites and bridged by two μ-$[P(SiMe_3)_2]^-$ or μ-$[As(SiMe_3)_2]^-$ ligands, respectively. In this way, a central, almost square $Mn_2E_2$ unit (E = P, As) is formed, leading to a roughly $C_{2v}$-symmetric $\{CpMnE_2\}$ coordination sphere with approximate $D_{2h}$ symmetry of the molecule. The average Mn–E bond length is 2.5099 Å in $[CpMn\{\mu\text{-}P(SiMe_3)_2\}]_2$ (**15**) and 2.5928 Å in $[CpMn\{\mu\text{-}As(SiMe_3)_2\}]_2$ (**16**). Whereas the average distance between manganese and the carbon atoms of the Cp ligand is 2.405 Å and 2.38 Å in the phosphorus and the arsenic derivative, respectively, both fall in the range observed for high-spin $Mn^{II}$ cyclopentadienides.[68] The Mn···Mn distances in **15** and **16** are, at 3.429(2) Å and 3.641(1) Å, respectively, not in the range of 2.170-3.291 Å normally found for manganese-manganese single bonds in the Cambridge Structural Database (CSD),[69] what makes the occurrence here very unlikely.

To investigate the magnetic properties of these pnictogen bridged dimanganese compounds, the molar magnetic susceptibility ($\chi_M$) was determined between 2 and 300 K with an applied field of H = 1000 G on polycrystalline, analytically pure samples (Figure 9).

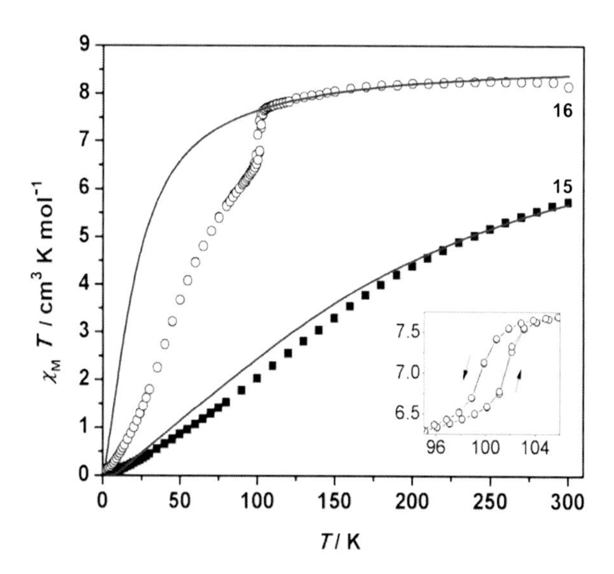


**Figure 9**: Temperature dependence of $\chi_M T$ for compounds **15** (squares) and **16** (circles). The red lines represent a theoretical fit of the experimental data. Inset: hysteresis loop for **16**.

For $[CpMn\{\mu\text{-}P(SiMe_3)_2\}]_2$ (**15**) a monotonic decrease of $\chi_M T$ can be observed: beginning with 5.71 $cm^3 K mol^{-1}$ at 300 K–about seventy-five per-cent of 8.75 $cm^3 K mol^{-1}$, the value predicted for two HS $^{MnII}$ centres–and ending with 0.027 $cm^3 K mol^{-1}$ at 2 K for a diamagnetic ground state. This behaviour indicates an antiferromagnetic exchange and the Heisenberg-Dirac-VanVleck (HDVV) model[70] with the spin Hamiltonian $H = -2J(S_{MnA} \cdot S_{MnB})$ fits the experimental data above 210 K and reveals a coupling constant of $J = -13.5\ cm^{-1}$ using $g = 2$ and $S_{Mn} = 5/2$. Surprisingly, this model does not well describe the temperature dependence of the susceptibility at temperatures lower than 210 K, but rather overestimates $\chi_M T$ until the diamagnetic ground state is reached. A possible explanation for this observation could be a spin crossover process, that is somehow masked by the increasing influence of the antiferromagnetic coupling and hence no quantitative conclusion can be drawn.

The susceptibility data measured for $[CpMn\{\mu\text{-}As(SiMe_3)_2\}]_2$ (**16**) differs from that found for the phosphorus analogue and is quite unusual. It starts with a monotonic decrease of $\chi_M T$ from 8.20 $cm^3 K mol^{-1}$ at 300 K to 7.67 $cm^3 K mol^{-1}$ at 105 K indeed, which can be fitted by the said Hamiltonian formalism to give a very small coupling constant of $J = -1.5\ cm^{-1}$ with $g = 2$ and $S_{Mn} = 5/2$ for two high-spin $Mn^{II}$ ions. And it arrives as well at 0.12 $cm^3 K mol^{-1}$ at 2 K–the diamagnetic ground state. At temperatures lower than 105 K, however, and before the diamagnetic ground state is reached, the model very much overestimates the experimental data. The observed abrupt decrease of the susceptibility between 105 and 96 K, can be explained by a spin crossover event. It even discloses a hysteresis loop, constant upon several cooling-warming cycles with a temperature sweep rate of 2 $K min^{-1}$. Additionally, the assumption that one of the two manganese ions undergoes a spin crossover process from the high spin (HS) with $S = 5/2$ to an intermediate spin state (IS) with $S = 3/2$ matches $\chi_M T$ at 96 K (6.37 $cm^3 K mol^{-1}$), that is around eighty per-cent of the value measured at 105 K. The result is a dimeric complex with one HS manganese(II) and one IS manganese(II) centre: $[CpMn^{HS}(\mu\text{-}As(SiMe_3)_2)_2Mn^{IS}Cp]$ or $[CpMn^{IS}(\mu\text{-}As(SiMe_3)_2)_2Mn^{HS}Cp]$. But the final diamagnetic state at 2 K requires both manganese ions to possess the same number of unpaired electrons and therefore to be in the same spin state. Hence, another spin crossover has to occur at the remaining HS manganese centre, which could take place below 75 K where there is again a rapid decrease of the magnetic susceptibility. This leads to two antiferromagnetically coupled manganese(II) ions in the intermediate spin state of $S = 3/2$ as the result of a thermally induced two-step spin crossover.

This consideration of the susceptibility data on complex **16** might help to better understand the behaviour of **15**: $\chi_M T$ of the phosphorus derivative at 300 K is with 5.71 $cm^3 K mol^{-1}$ quite similar to 6.37 $cm^3 K mol^{-1}$, the value found for the arsenide-bridged compound at 96 K. An analogous spin crossover to [CpMn$^{HS}$(μ-P(SiMe$_3$)$_2$)$_2$Mn$^{IS}$Cp] could be an explanation to that, as well as would be given by the one order of magnitude stronger antiferromagnetic coupling. Though both reasons are possible, the latter is the more likely one, because the experimental data obtained for temperatures above 250 K can be fitted well by a simple model for two exchange coupled HS Mn$^{II}$. The higher coupling constant $J$ of **15** might be due to a stronger ligand field of the bridging phosphides (i.e. more extensive overlap of metal-based magnetic orbitals with the ligand orbitals), the slightly different Mn–E–Mn angles might play an additional role in the mediation of the magnetic exchange. A manganese–manganese bond like in the diamagnetic Mn$^{II}$ phosphide complex [CpMn($^t$BuPH$_2$)(μ-$^t$BuPH)]$_2$,[71] though, cannot be excluded as a cause for the smaller than predicted $\chi_M T$ value by the susceptibility data alone. Yet, as already mentioned, the distance between the Mn centres is with 3.429(2) Å very large.

The exchange coupling constants determined for **15** and **16** are with $J$ = -13.5 cm$^{-1}$ and $J$ = -1.5 cm$^{-1}$ not as high as those calculated for the Mn$_5$P$_7$ cage compound of von Hänisch,[66] for which $J$ values up to -220 cm$^{-1}$ ($H = -J\, S_A \cdot S_B$) were reported. But it seems that the weak anti-ferromagnetic coupling in the arsenide bridged complex **16** in particular makes the observed spin crossover to intermediate spin states ($S$ = 3/2) possible. Spin crossover phenomena are uncommon for Mn$^{II}$ compounds and though gradual SCO processes have been accounted for manganocene derivatives, **16** is the only example that shows a two-step spin crossover with a sharp first transition that additionally shows hysteresis.

Another electron-deficient metallocene is chromocene (**17**), and might therefore show analogous reactivity to that of manganocene towards the substitution of Cp ligands for strong nucleophiles.

In the literature, diamagnetic polynuclear chromium complexes with bridging ligands bearing phosphorus or arsenic as donor atoms are well known, but their paramagnetic relatives are rare (Scheme H). In general, the magnetic exchange in chromium(II) complexes has not been studied extensively,[72] as Cr$^{II}$ ions tend more to the formation of chromium-chromium bonds. An exception are the dinuclear chromium(II) compounds reported by Fryzuk *et al.*, therein the metal centres are linked by hydride or chloride ligands.[73] As well as the triple-decker complex [{η$^5$-Cp*)Cr}$_2$(μ:η$^5$:η$^5$-P$_5$)]$^+$ with a *cyclo*-[P$_5$]$^-$ middle deck first found by Scherer *et al.*[74] that has stayed an object of interest until quite recently, because of its spin crossover behaviour.[75]

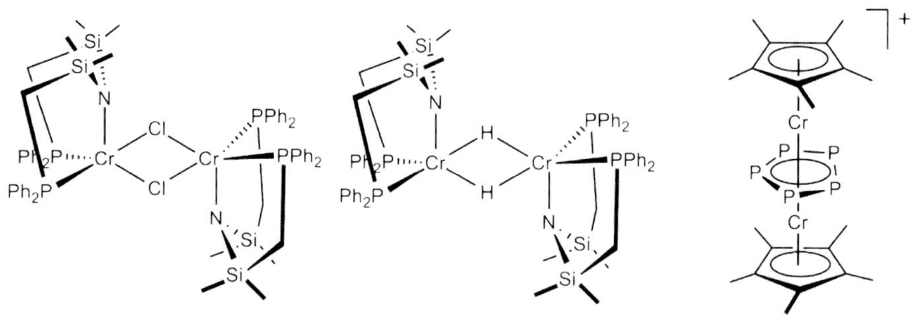

Fryzuk *et al.* **1994**[48]

Scherer *et al.* **1986**[49]
Hughes *et al.* **1994**,
Goeta *et al.* **2007**[50]

Scheme H: Selected paramagnetic dichromium(II) complexes.

The observation that chromium(II) complexes have mostly been investigated because of their interesting coordination behaviour, rather than because of their magnetic properties, made the exploration of pnictogen-bridged chromium complexes intriguing.

Reacting $Cp_2Cr$ (**17**) with $LiN(SiMe_3)_2$ in toluene at room temperature leads to the formation of the one-dimensional coordination polymer $^{1}_{\infty}[(\mu:\eta^2:\eta^5\text{-Cp})Cr\{\mu\text{-}N(SiMe_3)_2\}_2Li]$ (**18**) (Equation 9). This might happen because of the harder character according to the "hard and soft acids and bases" principle (HSAB) of the amide ligand compared to the rather soft phosphide and arsenide, as the outcome of the reaction is the same using a 1:1 or a 1:2 stoichiometry. The amide ligand seems to prefer a coordination to the harder lithium ion than solely to the softer chromium(II).

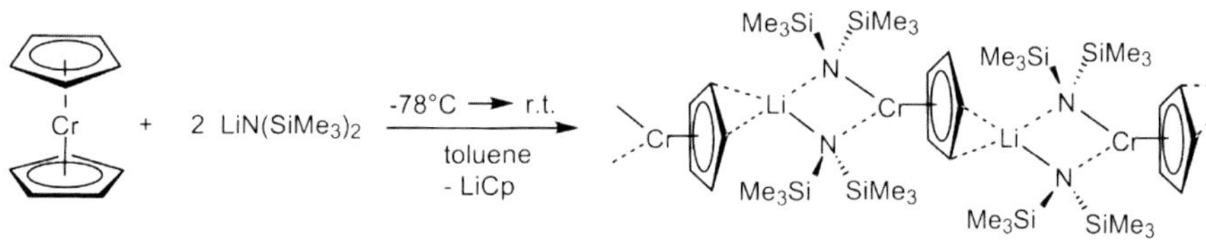

Cr + 2 $LiN(SiMe_3)_2$ —(-78°C → r.t., toluene, - LiCp)→ (9)

**17** **18** 21%

In contrast, the usage of the heavier lithiated trimethylsilyl-pnictenides $LiP(SiMe_3)_2$ or $LiAs(SiMe_3)_2$ results in the expected dichromium complexes $[(\eta^5\text{-Cp})Cr\{\mu\text{-}P(SiMe_3)_2\}]_2$ (**19**) and $[(\eta^5\text{-Cp})Cr\{\mu\text{-}As(SiMe_3)_2\}]_2$ (**20**), respectively (Equation 10). After filtering off precipitated LiCp, crystals for X-ray structure analysis can be obtained from saturated toluene solutions at -28 °C.

$$2\ \underset{\mathbf{17}}{[Cp_2Cr]} + 2\ LiE(SiMe_3)_2 \xrightarrow[\text{toluene}]{-78^\circ C \rightarrow r.t.} [CpCr\{\mu\text{-}E(SiMe_3)_2\}]_2 + 2\ LiCp \qquad (10)$$

**19** E = P, 32 %
**20** E = As, 45 %

Compound **18** crystallises as dark purple plates in the monoclinic space group $P2_1/n$. The asymmetric unit contains two independent molecules of $[(\mu{:}\eta^2{:}\eta^5\text{-}Cp)Cr\{\mu\text{-}N(SiMe_3)_2\}_2Li]$. The crystal structure analysis reveals that the chromium centre is five-coordinate by virtue of one $\eta^5$-Cp ligand and two $[N(SiMe_3)_2]^-$ ligands, hence the metal centre has a formal fourteen valence-electron count. The trimethylsilylamide ligands additionally coordinate a lithium ion to give a slightly asymmetric (Li–N1 2.079(4) Å, Li–N2 2.065(4) Å) diamond-shaped $CrN_2Li$ four-membered ring as the central unit of the monomer. Furthermore, $\eta^2$.coordination of the lithium by the Cp ring of the next molecule leads to the formation of a one-dimensional polymer, whereby the average $C_{Cp}$–Li distance is 2.624(1) Å, in contrast to 3.399(4)-3.769(6) Å for the remaining three carbon atoms of the Cp ring (*cf.* Figure 10).

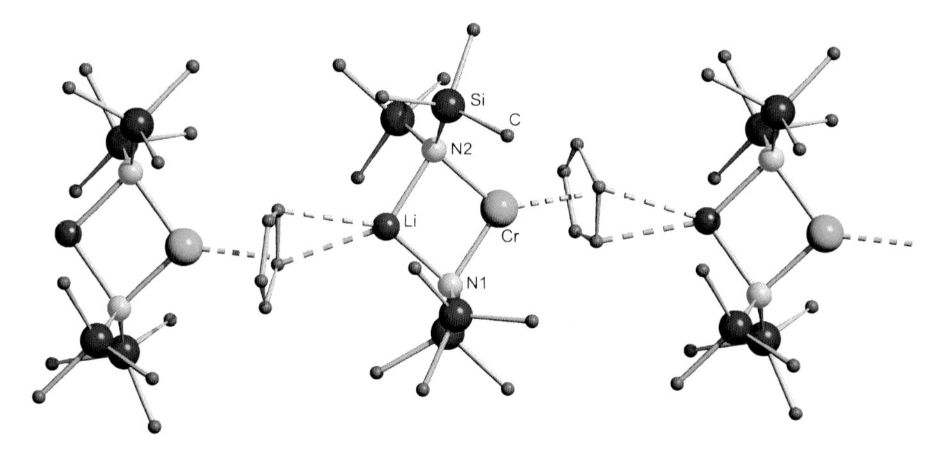


**Figure 10:** Molecular structure of **18** in the crystal. For clarity reasons, hydrogen atoms are omitted and carbon atoms are represented with reduced radii. Selected bond lengths [Å] and angles [°]: a) Cr–N1 2.093(1), Cr–N2 2.103(1), Li–N1 2.079(4), Li–N2 2.065(4), Cr···Li 2.761, N1–Cr–N2 96.28(7), Cr–N1–Li 82.9(1), Cr–N2–Li 83.0(1); b) Cr–N1 2.101(1), Cr–N2 2.084(1), Li–N1 2.055(4), Li–N2 2.078(4), Cr···Li 2.755, N1–Cr–N2 96.22(7), Cr–N1–Li 83.1(1), Cr–N2–Li 83.0(1);

By dissolving **18** in organic solvents, it is not very likely that the polymeric structure is preserved in solution. The $^1$H NMR spectrum (toluene-$D_8$) only shows a broad resonance at $\delta$ = 23.41 ppm that indicates the presence of the Cp ligand, as well as two signals at $\delta$ = 1.55 (sharp) and 2.85 ppm (broad) probably belonging to the trimethylsilyl groups. Nevertheless, the recorded $^7$Li NMR spectrum (toluene-$d_8$) reveals a signal at $\delta$ = 140.3 ppm, which is significantly downfield-shifted compared to the reported chemical shifts of $\delta \approx \pm 1$ ppm for $[LiN(SiMe_3)_2]$ in various solvents at various concentrations.[76] In addition to this, the broad and weak appearance of the signal indicates that the lithium containing species present in solution might be paramagnetic.

Compounds **19** and **20** both crystallise in the triclinic space group $P\bar{1}$, **19** with two independent molecules in the asymmetric unit, **20** with one. The shape of the dark red crystals of **19** is more rod-like, of **20** it resembles prisms.

As the crystal structure analysis shows (*cf.* Figure 11), both compounds are very similar: they bear two chromium centres that are each coordinated by one Cp ligand and bridged by two $[E(SiMe_3)_2]^-$ ligands (with E = P in **19** and As in **20**) to form an alternating $Cr_2E_2$ four-membered ring.

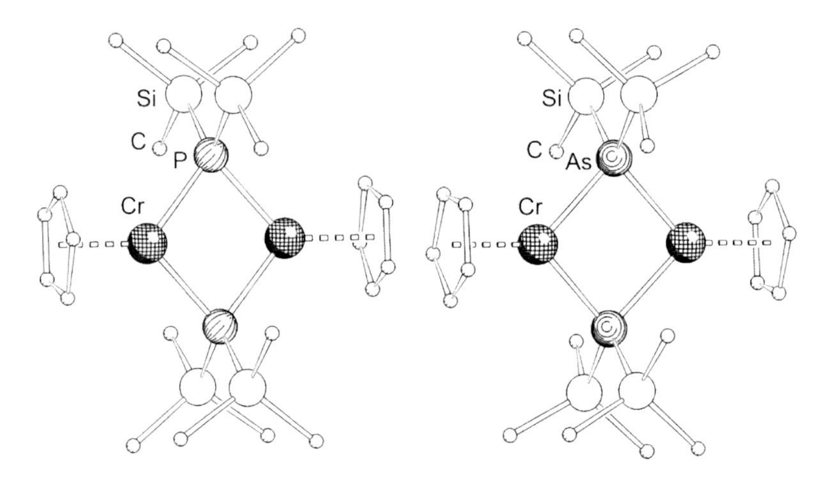


**Figure 11:** Molecular structure of **21** (left) and **22** (right) in the crystal. For clarity reasons, hydrogen atoms are omitted and carbon atoms are represented with reduced radii. Selected bond lengths [Å] and angles [°]: **21** a) Cr1–P1 2.5075(5), Cr1–P1A 2.5123(5), Cr1·· Cr1A 3.429, P1–Cr1–P1A 93.83(2), Cr1–P1–Cr1A 86.17(2); b) Cr1–P1 2.5017(5), Cr1–P1A 2.5197(5) Cr1···Cr1A 3.431, P1–Cr1–P1A 93.82(2), Cr1–P1–Cr1A 86.18(2); **22** Cr1–As1 2.5877(7), Cr1–As2 2.5980(8), Cr1··· Cr1A 3.641, As1–Cr1–As1A 90.81(2), Cr1–As1–Cr1A 89.20(2).

The $Cr^{II}$ ions have formal 14 valence electrons and are five-coordinate, with the Cp ligand occupying three coordination sites. This leads to approximate $C_{2v}$ symmetry regarding the $\{CpCrE_2\}$ unit and $D_{2h}$ symmetry for the whole molecule, where the Cp ligands are regarded as 'point ligands'. Thus, the structures of $[CpCr\{\mu\text{-}P(SiMe_3)_2\}]_2$ (**19**) and $[CpCr\{\mu\text{-}As(SiMe_3)_2\}]_2$ (**20**) are isotypical to the aforementioned dimanganese compounds **15** and **16**.

The Cr–C distances to the $\eta^5$-coordinated cyclopentadienyl ligands are between 2.27(1) and 2.33(2) Å in **19** and between 2.27(1) and 2.33(1) Å in **20**. In the phosphorus derivative, the average Cr–P bond length is 2.3839(9) Å, which lies well within the range of 2.2544-2.763 Å that can be found in the CSD for $\{Cr_2(\mu\text{-}P)_2\}$ structural motifs.[69] In the arsenide bridged compound, the average Cr-As bond length is 2.4978 Å, that results in a much longer separation of the two chromium centres (3.394(1) Å) than in the phosphide bridged one (3.066(1) Å). These chromium-chromium distances lie within 1.858–3.471 Å, the range for Cr–Cr bonds according to the CSD.[69] However, the Cr–Cr distance in **20** is very close to the upper limit, hence a Cr–Cr bond is not probable herein. The possibility of a bonding interaction of the two chromium centres in **19** cannot be excluded based solely on the crystallographic data. Nevertheless, the measurements of the magnetic susceptibility–whose results will be described below–indicate that the Cr ions only interact via exchange coupling through the bridging phosphide ligands.

On polycrystalline samples of **18**, **19** and **20**, magnetic susceptibility measurements at temperatures between 300 and 2 K with an applied field of 1000 G have been carried out, as well as isothermal magnetisation versus field measurements (for **19** and **20** see: Figure 12).

The magnetic susceptibility of $^1_\infty[(\mu;\eta^{2:5}\text{-}Cp)(Cr\{\mu\text{-}N(SiMe_3)_2\}_2Li]$ (**18**) remains independent of the temperature at a value of $\chi_M T$ = 2.80 cm$^3$ K mol$^{-1}$ until a slight decrease below 10 K to 2.35 cm$^3$ K mol$^{-1}$ is observed, which probably occurs due to zero-field splitting. The same result can be obtained at an applied field of 5000 G. Additionally, the field dependence of the magnetisation, i.e. *M(H)* has been measured with field strengths in the range of 0-7 T. Both isotherms, at 1.8 and 4 K, respectively, reach a saturation at approximately $M$ = 3.5 μB, which is in good accordance with four unpaired electrons at the chromium(II) centre. The isothermal magnetisation as well as the susceptibility data can well be fitted by a spin Hamiltonian ($H = -2J(S_{CrA} \cdot S_{CrB})$ for $Cr^{II}$ with a spin of $S = 2$, $g = 1.93$ and an axial zero-field splitting parameter of $D = -1.8$ cm$^{-1}$. These $g$- and $D$-values are consistent with those ones already reported for chromium(II) compounds.[77]

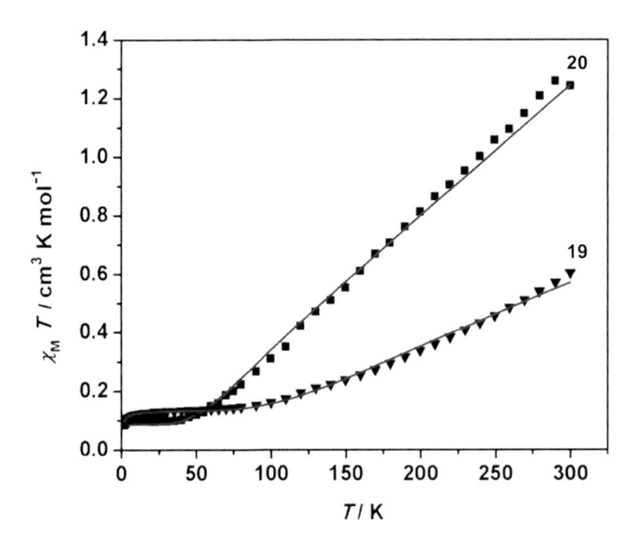


**Figure 12:** Temperature dependence of $\chi_M T$ for compounds **19** (triangles) and **20** (squares). The red lines represent a theoretical fit of the experimental data.

The dichromium complexes $[CpCr\{\mu\text{-}P(SiMe_3)_2\}]_2$ (**19**) and $[CpCr\{\mu\text{-}As(SiMe_3)_2\}]_2$ (**20**) show qualitatively the same temperature dependence of the magnetic susceptibility: Their $\chi_M T$ value at 300 K is with 0.60 $cm^3$ K $mol^{-1}$ for the phosphide and 1.24 $cm^3$ K $mol^{-1}$ for the arsenide derivative significantly smaller than the predicted spin-only value of 6.0 $cm^3$ K $mol^{-1}$ for two non-interacting $Cr^{II}$ ions ($g$ = 2). The susceptibility of both compounds then decreases continuously with temperature until it reaches $\chi_M T \approx 0$ at 2 K. In this way, a strong antiferromagnetic exchange between the chromium(II) centres leads to a diamagnetic ground state with $S = 0$ at very low temperatures. Small amounts of a paramagnetic impurity can possibly be accounted for the fact that $\chi_M T$ is only approximately zero at 2 K.

The experimental data was fitted with a HDVV model[70] for dimers bearing two metal ions with a spin of $S = 2$, including a $(T - \vartheta)$ term with $\vartheta$ being the Weiss constant to reflect intermolecular magnetic exchange. Using the same spin Hamiltonian formalism $H = -2J(S_{CrA} \cdot S_{CrB})$ as for the analogous manganese complexes **15** and **16**,[78] very good fits can be obtained with $g$ = 2 and $S_{Cr}$ = 2 for both chromium compounds. The model reveals a very large coupling constant of $J$ = -166 $cm^{-1}$ with $\vartheta$ = -0.7 K and $\rho$ = 0.022 (proportion of the paramagnetic impurity) for the phosphorus bridged compound **19**, and a still large $J$ value of -77.5 $cm^{-1}$ with $\vartheta$ = -0.4 K and $\rho$ = 0.014 for the arsenic bridged compound **20**. The nature of the minor paramagnetic impurity is unclear, but was modelled as a monomeric $Cr^{II}$ species. The parameters arising from this assumption produced good fits for magnetic susceptibility as well as the isothermal magnetisation versus field plots for **20**, which makes an oxidation product fairly unlikely.

Though both compounds show antiferromagnetic exchange and therefore a similar temperature dependent behaviour of the magnetic susceptibility, the strength of the exchange coupling is quite different. As the arsenide ligands already mediate a strong magnetic exchange ($J$ = -77.5 $cm^{-1}$), the exchange coupling constant reached by the phosphide ligands ($J$ = -166 $cm^{-1}$) is even more than twice as high. An important prerequisite for a strong superexchange is a good overlap of the valence orbitals involved. Given that the Cr–As bond is about 0.11 Å longer than the Cr–P bond, and that the Cr–As–Cr angle is 5-6° wider than the Cr–P–Cr one, which results in a 0.33 Å longer Cr···Cr distance in the arsenic bridged compound, it is conceivable that these variations of the structural parameters are responsible for the different strength of the antiferromagnetic exchange. Similar effects were observed for the manganese compounds **15** and **16**. Nevertheless, the antiferromagnetic coupling of both the phosphide and the arsenide complex is significantly stronger than the $J$-values of a few wavenumbers that can be achieved in the primarily studied polynuclear oxide-bridged chromium(III) compounds.[79] As already mentioned, there are only few examples of exchange coupled $Cr^{II}$ compounds. In case of the probably antiferromagnetically coupled triple-decker complex $[\{\eta^5\text{-Cp*})Cr\}_2(\mu{:}\eta^5{:}\eta^5\text{-}P_5)]^+$, unfortunately a determination of the coupling constant was hampered by a spin-crossover process.[75] A good comparison are the five-coordinate dichromium compounds $[(Ph_2PCH_2SiMe_2)_2NCr(\mu\text{-}Cl)]_2$ and $[(Ph_2PCH_2SiMe_2)_2NCr(\mu\text{-}H)]_2$: their two chromium(II) centres are bridged either by chloride ligands, which leads to an antiferromagnetic coupling with $J$ = -12.4 $cm^{-1}$, or by hydride ligands, resulting in a very strong antiferromagnetic coupling with $J$ = -139 $cm^{-1}$. The ability of hydride ligands to mediate a strong magnetic exchange manifests in the mixed-valent cage compound $[(\eta^5\text{-}C_5Me_4Et)_4Cr^{III}{}_3Cr^{II}(\mu\text{-}H)_5(\mu_3\text{-}H)_2]$, which shows that strong an antiferromagnetic exchange that the effective magnetic moment becomes temperature-independent.[80] However, the high coupling constants observed in our dichromium compounds reveal the potential of the heavier pnictogens, and particularly phosphorus, as donor atom for magnetic superexchange.

So far, it has been shown that varying the bridging pnictogen ligand can influence the magnetic properties of the investigated coordination complexes. Meanwhile, not much attention has been paid to the effects of the cyclopentadienyl ligands, which merely complete the coordination sphere of the metal ions. The described substitution of a Cp ligand for a bistrimethylsilylpnictenide had a great effect on the spin configuration of the chromium(II) ions: the described compounds **18**, **19** and **20** show a high spin configuration with $S = 2$, in contrast to the low spin $Cr^{II}$ in $Cp_2Cr$ where $S = 1$.[81] As the properties of cyclopentadienyl complexes of chromium(II) are significantly affected by ligand substitution,[82] the remaining Cp ligands seemed to be a good target for the introduction of ligands that counteract the weakening of the ligand field. The Cp*-analogues $[Cp^*CrE(SiMe_3)_2]_2$ (E = P, As) were thought to be easy to synthesise on the one hand, and to show a larger ligand field splitting due to the more strongly donating Cp* ligands that could enable a spin-crossover from HS to LS.

In contrast to $Cp_2Cr$ (**17**), $Cp^*_2Cr$ did not readily substitute one of the cyclopentadienyl ligands when treated with $LiE(SiMe_3)_2$. Along these lines an alternative synthesis was needed, which involved a two-step salt metathesis with anhydrous chromium(II) chloride as the metal source (Equation 11).

$$CrCl_2 + LiCp^* + LiE(SiMe_3)_2 \xrightarrow[\text{thf} \quad - E(SiMe_3)_3,\ - LiCl]{-78\,°C \rightarrow r.t.} \mathbf{21}\ (E = P,\ 33\ \%) \text{ or } \mathbf{22}\ (E = As,\ 56\ \%) \qquad (11)$$

[Cr] = Cp*Cr

Hence, $CrCl_2$ was reacted at -78 °C with one equivalent of LiCp* in thf to form $[Cp^*Cr(\mu\text{-}Cl)]_2$ in situ. The clear blue solution, which developed upon warming to room temperature, was then cooled to -78 °C, one equivalent of $LiE(SiMe_3)_2$ (E = P, As) added dropwise and warmed again to room temperature. From a filtered, saturated toluene solution, crystals of $[(\eta^5\text{-}Cp^*Cr)(\mu_3\text{-}P)]_4$ (**21**) and $[(\eta^5\text{-}Cp^*Cr)_3(\mu_3\text{-}As)_2]$ (**22**) could be obtained at -28 °C. However, these compounds are rather products of an oxidation reaction–with chromium no longer in the formal oxidation state +II, but formally +IV in **21** and formally +III in **22**–than of a sole salt elimination. The corresponding oxidising agents or reduction products, respectively, as well as the reaction mechanisms could not be clarified, as NMR studies were hindered by the paramagnetism of the compounds. A possible route, though, may be the formation of $[Cp^*Cr(\mu\text{-}Cl)]_2$ in the first step (blue solution), followed by a salt metathesis to $[Cp^*Cr\{\mu\text{-}E(SiMe_3)_2\}]_2$ (E = P, As)–in analogy with the observed stable Cp derivatives $[CpCr\{\mu\text{-}E(SiMe_3)_2\}]_2$ (E = P, As)–and dechloro-silylation, which leads on the one hand to the formation of $Me_3SiCl$ and on the other hand to a re-arrangement of the dimers to saturate the now 'naked' pnictogen atoms. As to the different oxidation states of chromium in **21** and **22**, an explanation might be provided by regarding the energies of the valence orbitals: those of $Cr^{IV}$ are thought to be on a lower energy level than those of $Cr^{III}$, and might therefore overlap better with the valence orbitals of phosphorus, whereas the higher in energy valence orbitals of arsenic are a better match for those of chromium(III).

Compound **21** crystallises as violet octahedra in the tetragonal space group $I\bar{4}$. X-ray structure analysis reveals a cube-like cage with chromium and phosphorus atoms alternately occupying the corners (P–Cr–P 98.32(3)-98.59(3)°), rather deriving from a chromium tetrahedron whose triangular surfaces are each capped by a naked phosphorus atom (Figure 13). The Cr···Cr distances therein are 2.931(2) and 2.935(1) Å, the P···P separation is 3.431(1) Å and all Cr–Cr–Cr and P–P–P angles are 60.0° (within standard deviations). An additional Cp* ligand completes the $\{Cp^*CrP_3\}$ piano stool coordination sphere of each chromium centre, with average distances of Cr–C 2.243 Å and Cr–P 2.622 Å.

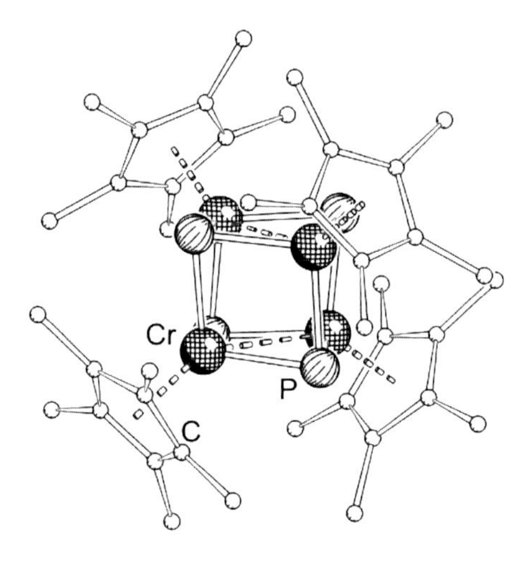


**Figure 13:** Molecular structure of **21** in the crystal. For clarity reasons, hydrogen atoms are omitted and carbon atoms are represented with reduced radii. Selected bond lengths [Å] and angles [°]: Cr1–P1 2.2609(9), Cr1–P1A 2.2646(8), Cr1–P1B 2.2610(9), Cr1–Cr1A 2.931(2), Cr1–Cr1B 2.935(1), P···P 3.431(1), Cr1A–Cr1–Cr1B 59.96(1), P1A–P1–P1B 59.81(2), P1–Cr1–P1A 98.59(3), P1–Cr1–P1B 98.32(3), P1A–Cr1–P1B 98.59(3).

Compound **22** crystallises in the orthorhombic space group *Pbca* as dark green blocks. X-ray structure analysis shows a trigonal bipyramid with a $\{(\eta^5\text{-}Cp^*Cr)_3\}$ unit as base and two $\mu_3$-arsenide ligands as tips (Figure 14). The average distance is 2.7735 Å between the chromium atoms and 2.4319 Å between chromium and arsenic. Within the $\{Cr_3\}$ triangle, the angles are 59.73(1)-60.48(1)°, and those between As–Cr–As are 97.48(1)-97.85(1)°. When assuming the Cp* ligands to undergo ring whizzing (Cr–C 2.224(2)-2.240(2) Å), an approximate $D_{3h}$ symmetry can be assigned to the molecule.

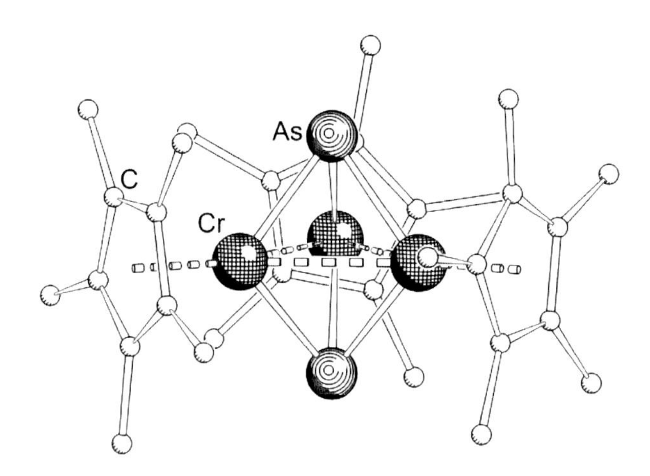


**Figure 14:** Molecular structure of **22** in the crystal. For clarity reasons, hydrogen atoms are omitted and carbon atoms are represented with reduced radii. Selected bond lengths [Å] and angles [°]: As1–Cr1 2.4304(4), As1–Cr2 2.4340(4), As1–Cr3 2.4381(4), As3–Cr1 (2.4258(4), As2–Cr2 2.4316(4), As2–Cr3 2.4315(4), Cr1–Cr2 2.7658(5), Cr1–Cr3 2.7869(4), Cr2–Cr3 2.7677(4), Cr1–Cr2–Cr3 60.48(1), Cr2–Cr3–Cr1 59.73(1), Cr3–Cr1–Cr2 59.79(1), Cr1–As1–Cr2 69.30(1), Cr1–As1–Cr3 69.84(1), Cr2–As1–Cr3 69.23(1), Cr1–As2–Cr2 69.41(1), Cr1–As2–Cr3 70.02(1), Cr2–As2–Cr3 69.37(1), As1–Cr1–As2 97.85(1), As1–Cr2–As2 97.59(1), As1–Cr3–As2 97.48(1).

However, the question, whether the spin density at the chromium centres is delocalised into metal-metal bonds or localised in the metal d-orbitals, cannot be answered by the observation alone that Cr–Cr distances (∅: 2.933 Å (**21**), 2.7735 Å (**22**)) lie within the range of 1.858-3.471 Å (∅: 2.755 Å), that can be found for metal-metal bonds in the CSD for chalcogen-bridged heterocubanes $[(CpM)_4(\mu_3\text{-}E)_4]^{n+}$ (M = Ti, V, Cr, Mo, Fe, Ru, Co, Ir; E = O, S; $n$ = 0, 1 ,2).[69] Insight into the electronic structure of **21** and **22** was therefore needed, hence measurements of the temperature dependence of the molar magnetic susceptibility $\chi_M T(T)$ and of the field dependence of the magnetisation $M()$ were carried out under the same conditions as for **18**, **19** and **20** (Figure 15).

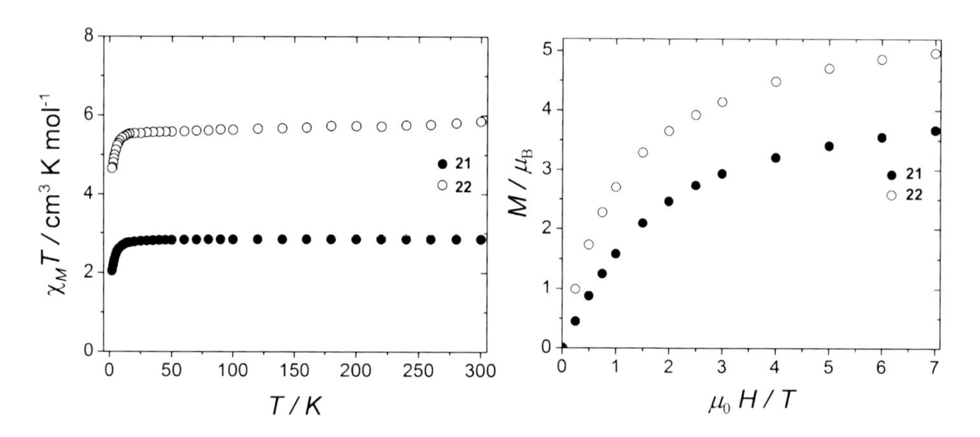


**Figure 15:** Temperature dependence of $\chi_M T$ in an applied field of 1 kG (left), and magnetisation ($M$) versus magnetic field ($H$) at 2 K (right), for **21** (shaded circles) and **22** (unshaded circles).

The molar magnetic susceptibility $\chi_M T$ of $[(\eta^5\text{-Cp*Cr})(\mu_3\text{-P})]_4$ (**21**) remains constant at 2.85 cm$^3$ K mol$^{-1}$ from 300 to 10 K, where a sharp drop to 2.05 cm$^3$ K mol$^{-1}$ can be observed. The magnetisation at 2 K shows a strong rise to 2.46 N$\mu_B$ at $H$ = 2.0 T, followed by a slower increase to 3.67 $\mu_B$ at 7.0 T without saturation. Though the values of $\chi_M T$(300 K) and $M$(7.0 T) are both much smaller than the 4.00 cm$^3$ K mol$^{-1}$ [70] and 8 N$\mu_B$, respectively, expected for four uncoupled chromium(IV) with $S$ = 1 ($g$ = 2), antiferromagnetic exchange should be very weak, due to the temperature independence of $\chi_M T$. Thus, an explanation for the magnetic behaviour has to take bonding interactions between the chromium ions into account. The predicted values for a $Cr^{IV}$ cluster with $S$ = 1 $\chi_M T$(300 K) = 3.00 cm$^3$ K mol$^{-1}$ [70] and $M$(7.0 T) = 4 N$\mu_B$, in contrast, reflect the measured data quite well. This quintet ground state is one of two possible spin configurations–$(a_1)^2$ $(e)^3(t_2)^3$ with $S$ = 2 or $(a_1)^2$ $(e)^4$ $(t_2)^4$ with $S$ = 1–that can be derived for a $Cr(IV)_4$ cubane from the molecular orbital (MO) scheme proposed for chalcogen-bridged cluster compounds $[(CpM)_4E_4]$ (E = O, S).[83]

To further underpin this assumption, density functional theory (DFT) calculations on the magnetic ground state of **21** were performed using the Gaussian suite of programs at the PBE/TZVP level. A quintet ground state was determined, with the triplet (S = 1; +20 kJ/mol), singlet (S = 0; +25 kJ/mol), septet (S = 3; +99 kJ/mol) and nonet (S = 4; +179 kJ/mol) lying above. Furthermore, in this quintet ground state, the spin density is almost completely located at the $Cr^{IV}$ centres according to the spin density plots and the Mayer bond orders for Cr–Cr bonds (0.22-0.24), supporting the presumption of bonding between the chromium ions in $[(\eta^5\text{-Cp*Cr})(\mu_3\text{-P})]_4$ (**21**).

In case of $[(\eta^5\text{-Cp*Cr})_3(\mu_3\text{-As})_2]$ (**22**), the magnetic susceptibility decreases slightly from 300 to 10 K and drops abruptly after that to reach $\chi_M T = 4.65\ \text{cm}^3\ \text{K mol}^{-1}$. The value found for the $\chi_M T$ term at 300 K, $5.85\ \text{cm}^3\ \text{K mol}^{-1}$, resembles well the $5.63\ \text{cm}^3\ \text{K mol}^{-1}$ predicted for an uncoupled system with three $Cr^{III}$ of S = 3/2, as does the magnetisation at 2 K with M(7.0 T) = 4.96 NμB compared to the theoretical value of 4.5.[84]

Additional DFT calculations (PBE/TZVP and PBE0/TZVP) substantiate a decet ground state with S = 9/2 with the other spin multiplicities being relatively higher in energy: quartet (S = 3/2; +55 kJ/mol), octet (S = 7/2; +66 kJ/mol), sextet (S = 5/2; +74 kJ/mol) and doublet (S = 1/2; +82 kJ/mol). The calculated Mayer bond orders for Cr–Cr bonds are much smaller in **22** than in **21** and lie between 0.07 and 0.10, indicating that a metal-metal bond is unlikely.

The formal oxidation state of +IV in **21** is not very common for chromium.[85] Accordingly, the oxidation from $Cr^{II}$ in the starting material to $Cr^{IV}$ in the cluster is a rather unusual feature, particularly because chromium(II) ions tend more to be oxidised to +III or +VI.[9] Furthermore, chromium(IV) compounds mostly bear harder oxygen or nitrogen ligands as donors,[86] instead of the relatively soft $P^{3-}$ and $Cp^{*-}$. In organometallic complexes especially, chromium(IV) is rare[84] and, as far as we know, no polynuclear $Cr^{IV}$ compound has been reported yet. Nevertheless, similar chromium(III) cluster compounds with $\{Cr_4E_4\}$ moieties are well known for E = O, S, Se.[87] There are, however, only five phosphorus-containing structurally analogue heterocubanes, two of them with cobalt,[88] one with tin[89], one with tungsten[90] and the just recently in our workgroup synthesised $[CpMnP]_4$.[91] Compound **21** is still the only example containing chromium, even if compounds having less similar structures like $[(CpV)_4(P_3)_2]$[92], $[(CpNi)_3(\mu_3\text{-P})(P_4)]$[93] and $[(CpFe)_4(P_2)_2]$[94] are taken into account.

A number of coordination cages like **22** with a central $M_3E_2$ motif are already known for different metals with E being one of the heavier pnictogen elements. But compound **22** is again the first cluster of this kind bearing chromium. To account the upcoming reactions, a more detailed insight into the literature will be given later on.

This synthetic method could not only serve as a way to unusual chromium(IV) compounds, but oxidative addition could be a general route to complexes with naked pnictogen ligands. Having in mind that iron is quite common in all three, hetero-cluster compounds (especially with S[95]), $P_n$ ligand complexes as well as magnetically active compounds, it was natural to turn to this metal to see, whether analogue compounds can be realised.

Along these lines, iron(II)chloride was reacted with LiCp* and $LiE(SiMe_3)_2$ (E = P, As) under the same conditions as described above for $CrCl_2$. Both reaction mixtures turned green after addition of LiCp*, and changed to a very dark brown when treated with $LiE(SiMe_3)_2$. Black crystals of $[(\eta^5\text{-}Cp^*Fe)_3(\mu_3\text{-}P)_2]$ (**23**) and $[(\eta^5\text{-}Cp^*Fe)_3(\mu_3\text{-}As)_2]$ (**24**), respectively, could be obtained after filtration from saturated solutions at -28 °C (Equation 12). NMR studies to clarify the reaction mechanism and to enable a future identification of the compounds have not been possible, due to both the paramagnetism and the high solubility, the latter making it impossible to wash reaction mixture residues off the isolated crystals properly.

$$FeCl_2 + LiCp^* + LiE(SiMe_3)_2 \xrightarrow[\text{thf} \quad -E(SiMe_3)_3,\ -LiCl]{-78\,°C \rightarrow r.t.} [Fe]_3E_2 \qquad (12)$$

(via $[Cp^*Fe(\mu\text{-}Cl)]_2$)

**23** E = P, 23 %
**24** E = As, 19 %
[Fe] = Cp*Fe

Compound **23** crystallises as black plates in the hexagonal space group *P6₅*, **24** as black blocks in the hexagonal space group *P6₁*. Crystal structure analysis revealed that both compounds consist of a triangular base of three Cp*-substituted iron centres that is capped by two $P^{3-}$ or $As^{3-}$, respectively, to form a trigonal bipyramid (Figure 16,Figure 17).

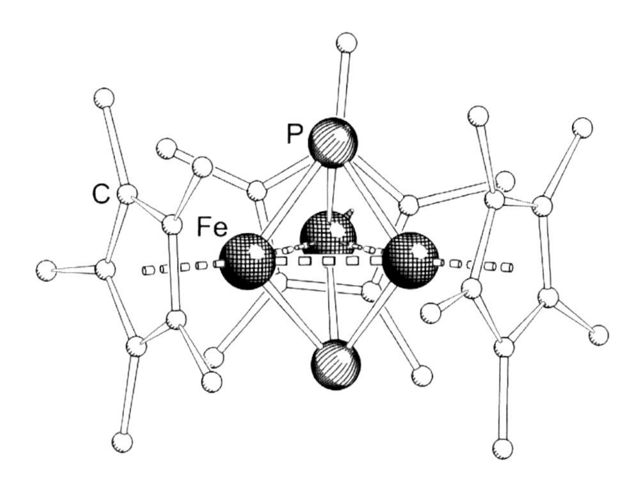


Figure 16: Molecular structure of 23 in the crystal. For clarity reasons, hydrogen atoms are omitted and carbon atoms are represented with reduced radii. Selected bond lengths [Å] and angles [°]:P1–Fe1 2.196(1), P1–Fe2 2.225(1), P1–Fe3 2.177(1), P2–Fe1 2.206(1), P2–Fe2 2.231(1), P2–Fe3 2.183(1), Fe1–Fe2 2.3981(8), Fe1–Fe3 2.5252(7), Fe2–Fe3 2.4592(8), Fe1–Fe2–Fe3 62.64(2), Fe2–Fe3–Fe1 57.50(2), Fe3–Fe1–Fe2 59.87(2), Fe1–P1–Fe2 65.70(3), Fe1–P1–Fe3 70.56(3), Fe2–P1–Fe3 67.91(3), Fe1–P2–Fe2 65.44(3), Fe1–P2–Fe3 70.25(3), Fe2–P2–Fe3 67.71(3), P1–Fe1–P2 99.62(4), P1–Fe2–P2 97.97(5), P1–Fe3–P2 100.92(4).

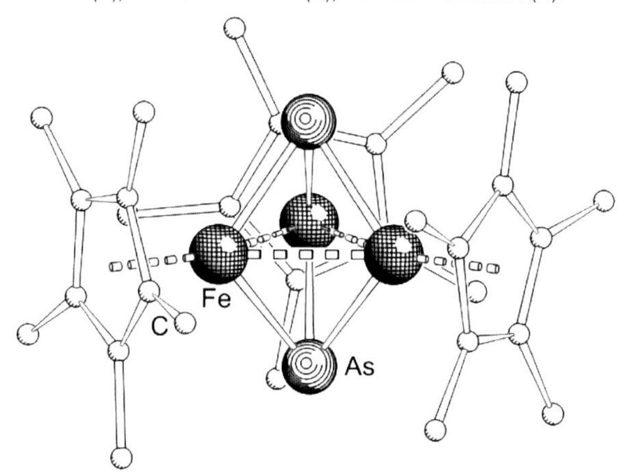


Figure 17: Molecular structure of 24 in the crystal. For clarity reasons, hydrogen atoms are omitted and carbon atoms are represented with reduced radii. Selected bond lengths [Å] and angles [°]: As1–Fe1 2.3345(7), As1–Fe2 2.3137(7), As1–Fe3 2.3222(6), As2–Fe1 2.3166(6), As2–Fe2 2.2936(7), As2–Fe3 2.3048(6), Fe1–Fe2 2.5579(8), Fe1–Fe3 2.5150(6), Fe2–Fe3 2.5993(8), Fe1–Fe2–Fe3 58.37(2), Fe2–Fe3–Fe1 59.99(2), Fe3–Fe1–Fe2 61.64(2), Fe1–As1–Fe2 66.77(2), Fe1–As1–Fe3 68.21(2), Fe2–As1–Fe3 68.21(2), Fe1–As2–Fe2 67.40(2), Fe1–As2–Fe3 65.94(2), Fe2–As2–Fe3 68.84(2), As1–Fe1–As2 99.98(2), As1–Fe2–As2 101.29(3), As1–Fe3–As2 100.70(2).

The mean bond distances between the iron centres and the naked pnictogen atoms are 2.203 Å in the phosphorus and 2.314 Å in the arsenic compound, with corresponding E–Fe–E angles of 99.5° (**23**) and 100.7° (**24**). Within the $(Cp^*Fe)_3$ subunit, the average Fe–Fe bond is 2.4608 Å in **23** and 2.5574 Å in **24** with all Fe–Fe–Fe angles approximately 60°. Interestingly, not only the Fe–F bonds are elongated in the arsenic compound compared to the phosphorus one, but also the iron-iron bonds (about 0.1 Å). This widening of the triangular base could possibly occur because the interaction of the iron centres and the P ligand is stronger than with the As ligand. In analogy to $[(\eta^5\text{-}Cp^*Cr)_3(\mu_3\text{-}As)_2]$ (**22**), compounds **23** and **24** have approximate $D_{3h}$ symmetry.
Unfortunately, the surely intriguing investigation of the magnetic properties has not been finished up to date, and can therefore not be presented within this thesis.

As already mentioned above, the trigonal bipyramide $\{M_3E_2\}$ is a common structural motif for heteroclusters. The largest group of such cages are iron compounds, in which the metal centres are saturated by carbonyl ligands and the P, As or Sb atoms, respectively, bear either organic substituents[96] or metal complex fragments.[97] $[Fe_3(CO)_9As_2]$,[98] $[Fe_3(CO)_9Bi_2]$[99] and $[Ru_3(CO)_9Bi_2]$,[100] however , contain naked $\{\mu_3\text{-}E\}^{3-}$ ligands. But they all contain iron(II) ions, in contrast to **23** and **24**, in which the iron centres are formally in the oxidation state +III. Nevertheless, cage compounds isostructural to **22**, **23** and **24**, with Cp ligands instead of CO ligands, are rare: The groups of Roesky and von Hänisch reported aluminium analogues–the former is the not yet structurally characterised $[(Cp^*Al)_3(\mu_3\text{-}Bi)_2]$,[101] the latter is $[(Cp^*Al)_3(\mu_3\text{-}As)_2]$.[102] In addition, the cobalt derivative $[(Cp''Co)_3(\mu_3\text{-}P)_2]$ ($Cp'' = C_5H_3{}^tBu_2$) was realised by Scherer *et al.*[88b]

Gathering all four presented cage compounds–$[(\eta^5\text{-}Cp^*Cr)(\mu_3\text{-}P)]_4$ (**21**), $[(\eta^5\text{-}Cp^*Cr)_3(\mu_3\text{-}As)_2]$ (**22**), $[(\eta^5\text{-}Cp^*Fe)_3(\mu_3\text{-}P)_2]$ (**23**) and $[(\eta^5\text{-}Cp^*Fe)_3(\mu_3\text{-}As)_2]$ (**24**)–the addition of $LiE(SiMe_3)_2$ (E = P, As) to an *in situ* generated $[Cp^*M(\mu\text{-}Cl)]_2$ (M = Fe, Cr) with an simultaneous oxidation of the metal centres seems to be a comfortable way to heteroatomic clusters with naked pnictogen atoms. The fact that halogen-bridged dimers of this kind are known for many transition metals[103] might bear the possibility to transfer this synthesis route to obtain a larger variety of pnictogen-containing cage complexes.
Moreover, the complexes themselves, with their un-substituted phosphorus and arsenic atoms, provide the opportunity to use these in further reactions. On the one hand, the successful

oxidation of the above mentioned cobalt compound with elemental chalcogens to [(Cp''Co)$_3$($\mu_3$-PQ)$_2$] (Q = O, S, Se) and [(Cp''Co)$_3$($\mu_3$-PQ)($\mu_3$-P)] (Q = Se, Te) shows that it should at least principally be possible to oxidise these clusters.[104] But on account of the high air sensitivity–solutions of the chromium compounds immediately turn green–, a milder oxygen reagent would be more appropriate. Introduction into the supramolecular chemistry, on the other hand, should be promising as well: Schauer *et al.* observed a one-dimensional polymer, in which {(Fe(CO)$_3$)$_3$($\mu_3$-P)$_2$}$^{2-}$ units are linked by gold ions.[105]

Apart from the mere synthetic point of view, these reactions are also interesting because they all lead to a decrease of electron density at the pnictogen ligands and the resulting effects on the magnetic properties would be worth studying.

# 4. Experimental Section

## 4.1 General remarks

### 4.1.1 Preparative procedures

All manipulations were performed using standard Schlenk and dry-box techniques under an atmosphere of dry dinitrogen or argon. Traces of oxygen and moisture were removed from the inert gas by passing them over BASF R 3-11 ($CuO/MgSiO_3$) catalyst, through concentrated $H_2SO_4$ and over coarsely granulated silica gel, in that order.

All solvents were degassed and distilled from appropriate drying agents under an atmosphere of dinitrogen prior to use. Boiling the solvents under reflux for at least four hours preceded the distillation process. Hexane and Pentane were distilled from Na/K alloy, toluene from Na, $Et_2O$, dme and thf from Na/benzophenone, $CH_2Cl_2$ from $CaH_2$ and $CH_3CN$ as well as *ortho*-dichlorobenzene from $CaCl_2$. The deuterated solvents $C_6D_6$ and pyridine-$d_5$ were degassed and destilled from Na or KOH, respectively. Both were additionally stored over molecular sieve (4 Å) which had previously been dried for two hours under high vacuum at 100 °C.

Diatomaceous earth was routinely stored at 110 °C prior to use, then dried in vacuum with the aid of a heat gun. Silica gel 60 (particle size: 0.063 – 0.2 mm) used for the column chromatography was heated under vacuum (3 d, $10^{-3}$ mbar, 230 °C) prior to use.

### 4.1.2 Starting materials

The following substances were available at our work group, kindly donated by the persons named in parenthesis or were prepared according to literature procedures: [Cp*Fe($\eta^5$-$P_5$)] (**1**),[29] K[P($SiMe_3$)$_2$],[36] [$FeBr_2$(dme)],[106] LiCp*, [Cp*Fe($\eta^5$-${}^iPr_3C_3P_2$)] (**5**),[39, 44b] NaCp''' (Dr. C. Schwarzmaier), NaCp$^{BIG}$ (Dr. S. Heinl), NaCp$^{4iPr}$ (Prof. Dr. H. Sitzmann, TU Kaiserslautern), [$Cp_2Mn$] (**14**), Li[P($SiMe_3$)$_2$],[107] Li[As($SiMe_3$)$_2$] (S. Bauer, C. Marquart, E.-M. Rummel), [$Cp_2Cr$] (**17**), $CrCl_2$;
White phosphorus was sublimed and stored under argon and light exclusion.
The following compounds were obtained from commercial suppliers: Adamantane ($C_{10}H_{16}$), CuCl, CuBr, CuI, $FeBr_3$, Li[N($SiMe_3$)$_2$], $FeCl_2$;

### 4.1.3 Characterisation methods

**Solution NMR spectra** were recorded at the NMR department of the University of Regensburg using a Bruker Avance 300 or 400 spectrometer. Samples were referenced against TMS ($^{1}H$), 1 M LiCl in $D_2O$ ($^{7}Li$), and 85% $H_3PO_4$ ($^{31}P$) as external standards. Chemical shifts are reported in ppm, according to the $\delta$-scale, the coupling constants $J$ in Hz.

**Solid-state MAS NMR spectra** were recorded by Prof. Dr. Werner Kremer (research group of Prof. Dr. Dr. H. R. Kalbitzer, University of Regensburg) on a Bruker Avance 300 solid state spectrometer. The NMR spectra were processed using the TopSpin 2.1 program (Bruker).

**LIFD MS and ESI MS spectra** were measured by the MS department of the University of Regensburg using a Finnigan MAT 95 (LIFD) and a ThermoQuest Finnigan TSQ 7000 (ESI) spectrometer. The identity of the observed fragments was assigned according to the mass/charge ($m/z$) ratio, the isotope pattern and comparison of the experimental signals with simulated ones, which were generated by the ChemDraw Ultra 10.0 software (Cambridge Soft).

**Elemental analyses** were performed at the micro analytical laboratory of the University of Regensburg on a Vario EL III instrument or obtained using the elemental analysis service of London Metropolitan University, U.K. (Mr. S. Boyer).

**IR spectra** were recorded on a VARIAN FTS-800 FT-IR spectrometer in solution.

**Magnetic susceptibility measurements** on polycrystalline samples were carried out using a Quantum Design MPMS-7 SQUID magnetometer at temperatures in the range 2-300 K. U.K. The experiments were performed and fitted by Dr. Floriana Tuna and M.Sc. Eufemio Moreno Pineda at the National EPR Facility, Photon Science Institute, The University of Manchester (UK).

**Yields** were determined from isolated crystals of the respective compound, if not mentioned otherwise. Calculation from NMR spectra is noted in parentheses.

### 4.1.4 Theoretical calculations

Joseph J. W. McDouall (School of Chemistry, The University of Manchester, UK) performed the DFT (density functional theory) calculations with the Gaussian 09 suite of programs[108] on the following compounds (used functional and basis sets are given in parentheses):
**21** (PBE[109] and TZVP[110]) and **22** (PBE or PBE0[111] and TZVP).

## 4.2 Synthetic details

### 4.2.1 $C_{10}H_{16}$@[{Cp*Fe($\eta^{1:1:1:1:1:5}$-$P_5$)}$_{12}$($CuCl$)$_{20-y}$] (2)

In a thin schlenk tube, a solution of 17 mg (0.17 mmol) CuCl in 2 mL acetonitrile is layered over a solution of 30 mg (0.09 mmol) [Cp*Fe($\eta^5$-$P_5$)] (**1**) and 10 mg (0.07 mmol) adamantane ($C_{10}H_{16}$) in 2 mL dichloromethane. After complete diffusion, brown blocks of $C_{10}H_{16}$@[{Cp*Fe($\eta^{1:1:1:1:1:5}$-$P_5$)}$_{12}$($CuCl$)$_{20-y}$] (**2**) are formed. The mother liquor is decanted, then the crystals are washed twice with 1 mL toluene and once very carefully 1 mL with dichloromethane. The solvent is removed under reduced pressure and the crystals are dried in high vacuum.

Analytical data for **2**:

| | |
|---|---|
| **Yield** | 31 mg (5.19 µmol, 68 %). |
| **$^1$H NMR** (pyridine-$d_5$) | δ [ppm] = 1.36 (s, 15H, $C_5\underline{Me}_5$), 1.65 (s, 12 H, $CH_2$), 1.81 (s, 4H, CH). |
| **$^{31}$P{$^1$H} NMR** (pyridine-$d_5$) | δ [ppm] = 139.95 (s, 5P, [Cp*Fe($\eta^5$-$\underline{P}_5$)]). |
| **$^1$H MAS NMR** | δ [ppm] = 1.7 (s, br, $\omega_{1/2}$ = 727 Hz, $C_5\underline{Me}_5$), -3.2 (s, $C_{10}\underline{H}_{16}$). |
| **ESI-MS** ($CH_2Cl_2$) | m/z [%] = 1596.3 [($Cp^*FeP_5)_3Cu_6Cl_5$]$^+$, 1496.3 [($Cp^*FeP_5)_3Cu_5Cl_4$]$^+$, 1396.5 [($Cp^*FeP_5)_3Cu_4Cl_3$]$^+$, 1298.5 [($Cp^*FeP_5)_3Cu_3Cl_2$]$^+$, 1052.6 [($Cp^*FeP_5)_2Cu_4Cl_3$]$^+$, 952.6 [($Cp^*FeP_5)_2Cu_3Cl_2$]$^+$, 854.7 [($Cp^*FeP_5)_2Cu_2Cl$]$^+$, 754.8 [($Cp^*FeP_5)_2Cu$]$^+$, 449.9 [($Cp^*FeP_5$)Cu($CH_3CN$)]$^+$, 408.8 [($Cp^*FeP_5$)Cu]$^+$. |
| **Elemental analysis** | calcd. for $C_{130}H_{196}Fe_{12}Cu_{17}Cl_{17}$ (5970.56 g mol$^{-1}$) C 26.15, H 3.31; found C 26.13, H 3.34; (y = 3). |

### 4.2.2 $C_{10}H_{16}$@[{Cp*Fe($\eta^{1:1:1:1:1:5}$-$P_5$)}$_{12}$($CuBr$)$_{20-y}$] (3)

In a thin schlenk tube, a solution of 25 mg (0.17 mmol) CuBr in 2 mL acetonitrile is layered over a solution of 30 mg (0.09 mmol) [Cp*Fe($\eta^5$-$P_5$)] (**1**) and 10 mg (0.07 mmol) adamantane ($C_{10}H_{16}$) in 2 mL dichloromethane. After complete diffusion, brown blocks of $C_{10}H_{16}$@[{Cp*Fe($\eta^{1:1:1:1:1:5}$-$P_5$)}$_{12}$($CuBr$)$_{20-y}$] (**3**) are formed. The mother liquor is decanted, then the crystals are washed twice with 1 mL toluene and once very carefully 1 mL with dichloromethane. The solvent is removed under reduced pressure and the crystals are dried in high vacuum.

Compound **3** can as well be obtained using toluene instead of dichloromethane, but the diffusion process takes longer, the yields are smaller and it crystallises in a different unit cell. The complete crystal structure analysis has only been performed on crystals grown from toluene/$CH_3CN$ because of their better quality.

Analytical data for **3**:

| | |
|---|---|
| **Yield** | 25 mg (3.49 µmol, 50 %) |
| $^{1}$**H NMR** (pyridine-$d_5$) | δ [ppm] = 1.34 (s, 15H, $C_5\underline{Me}_5$), 1.66 (s, 12 H, $CH_2$), 1.83 (s, 4H, CH). |
| $^{31}$**P{**$^{1}$**H} NMR** (pyridine-$d_5$) | δ [ppm] = 144.86 (s, 5P, [Cp*Fe($\eta^5$-$\underline{P}_5$)]). |
| $^{1}$**H MAS NMR** | δ [ppm] = 2.1 (s, br, $\omega_{1/2}$ = 2521 Hz, $C_5\underline{Me}_5$), -3.1 (s, $C_{10}\underline{H}_{16}$). |
| **ESI-MS** ($CH_2Cl_2$) | m/z [%] = 1042.5 [$(Cp^*FeP_5)_2Cu_3Br_2]^+$, 898.6 [$(Cp^*FeP_5)_2Cu_2Br]^+$, 754.8 [$(Cp^*FeP_5)_2Cu]^+$, 593.7 [$(Cp^*FeP_5)Cu_2Br(CH_3CN)]^+$, 552.7 [$(Cp^*FeP_5)Cu_2Br]^+$, 449.9 [$(Cp^*FeP_5)Cu(CH_3CN)]^+$, 408.8 [$(Cp^*FeP_5)Cu]^+$. |
| **Elemental analysis** | calcd. for $C_{130}H_{196}Fe_{12}Cu_{17}Br_{17}$ (6726.23 g mol$^{-1}$) C 23.22, H 2.92; found C 23.21, H 2.94; (y = 3). |

### 4.2.3 $C_6H_4Cl_2$@[{Cp*Fe($\eta^{1:1:1:1:5}$-$P_5$)}$_8$(CuI)$_{28}$($CH_3CN$)$_{10}$] (4)

In a thin schlenk tube, a solution of 17 mg (0.09 mmol) CuI in 10 mL of a 1:1-mixture of acetonitrile and *ortho*-dichlorobenzene is layered over a solution of 30 mg (0.09 mmol) [Cp*Fe($\eta^5$-$P_5$)] (**1**) in 5 mL *ortho*-dichlorobenzene. After one third of the solutions have merged, a few red plates of $C_6H_4Cl_2$@[{Cp*Fe($\eta^{1:1:1:1:1:5}$-$P_5$)}$_8$(CuI)$_{28}$($CH_3CN$)$_{10}$] (**4**) are formed at the wall of the schlenk tube in the boundary layer region. They are vanished, when about two thirds of the solutions have merged and crystals of $\mathbf{P_{2d}}$ and $\mathbf{P_{odCB}}$ emerge.

Analytical data for **4:**

| | |
|---|---|
| **Yield** | a few crystals |

### 4.2.4 $^{1}_{\infty}[\{Cp^{*}Fe(\eta^{1:1:5}\text{-}{}^{i}Pr_3C_3P_2)\}Cu_2(\mu\text{-}Cl)_2(CH_3CN)]$ (6)

A solution of 35 mg (0.35 mmol) CuCl in 1 mL acetonitrile is layered over a solution of 73 mg (0.17 mmol) $[Cp^{*}Fe(\eta^{5}\text{-}{}^{i}Pr_3C_3P_2)]$ (**5**) in 1 mL toluene in a thin schlenk tube. Overnight, a colourless precipitate forms. The reaction mixture is then both underlayered with one additional millilitre of toluene and overlayered with 1 mL acetonitrile. After two days, small orange crystals of $^{1}_{\infty}[\{Cp^{*}Fe(\eta^{1:1:5}\text{-}{}^{i}Pr_3C_3P_2)\}Cu_2(\mu\text{-}Cl)_2(CH_3CN)]$ (**6**) are formed. The mother liquor is decanted and the crystals are washed with $CH_2Cl_2$ to remove precipitated CuCl, then dried under vacuum.

Analytical data for **6**:

| | |
|---|---|
| **Yield** | 5 mg (7.6 µmol, 4 %) |
| **ESI-MS** ($CH_2Cl_2/CH_3CN$) | m/z [%] = 1616.4 $[\{Cp^{*}Fe({}^{i}Pr_3C_3P_2)\}_3Cu_4Cl_3]^{+}$, 1515.4 $[\{Cp^{*}Fe({}^{i}Pr_3C_3P_2)\}_3Cu_3Cl_2]^{+}$, 1097.2 $[\{Cp^{*}Fe({}^{i}Pr_3C_3P_2)\}_2Cu_3Cl_2]^{+}$, 999.3 $[\{Cp^{*}Fe({}^{i}Pr_3C_3P_2)\}_2Cu_2Cl]^{+}$, 899.4 $[\{Cp^{*}Fe({}^{i}Pr_3C_3P_2)\}_2Cu]^{+}$, 720.0 $[\{Cp^{*}Fe({}^{i}Pr_3C_3P_2)\}Cu_3Cl_2(CH_3CN)]^{+}$, 678.9 $[\{Cp^{*}Fe({}^{i}Pr_3C_3P_2)\}Cu_3Cl_2]^{+}$ 622.0 $[\{Cp^{*}Fe({}^{i}Pr_3C_3P_2)\}Cu_2Cl(CH_3CN)]^{+}$, 522.1 $[\{Cp^{*}Fe({}^{i}Pr_3C_3P_2)\}Cu(CH_3CN)]^{+}$, 480.2 $[\{Cp^{*}Fe({}^{i}Pr_3C_3P_2)\}Cu]^{+}$, 418.1 $[Cp^{*}Fe({}^{i}Pr_3C_3P_2)]^{+}$. |
| **Elemental analysis** | calcd. for $C_{266}H_{435}P_{24}Fe_{12}Cu_{24}Cl_{24}$ ($^{1}/_{12}$ eq $CH_3CN$) (7436.90 g $mol^{-1}$) C 41.21, H 5.90, N 0.20; found C 42.96, H 5.90, N 0.19. |

### 4.2.5 $^{1}_{\infty}[\{Cp^{*}Fe(\eta^{1:1:5}\text{-}{}^{i}Pr_3C_3P_2)\}Cu_2(\mu\text{-}Br)_2(CH_3CN)]$ (7)

A solution of 24 mg (0.16 mmol) CuBr in 1 mL acetonitrile is layered over a solution of 35 mg (0.08 mmol) $[Cp^{*}Fe(\eta^{5}\text{-}{}^{i}Pr_3C_3P_2)]$ (**5**) in 0.5 mL toluene in a thin schlenk tube. After diffusion, small orange platelets of $^{1}_{\infty}[\{Cp^{*}Fe(\eta^{1:1:5}\text{-}{}^{i}Pr_3C_3P_2)\}Cu_2(\mu\text{-}Br)_2(CH_3CN)]$ (**7**) are formed. The mother liquor is decanted and the crystals are washed with $CH_2Cl_2$ to remove precipitated CuBr, then dried under vacuum.

Analytical data for **7**:

| | |
|---|---|
| **Yield** | 5 mg (6.7 µmol, 8 %) |
| **ESI-MS** (toluene/$CH_3CN$) | m/z [%] = 1043.3 $[\{Cp^*Fe(^iPr_3C_3P_2)\}_2Cu_2Br]^+$, 899.3 $[\{Cp^*Fe(^iPr_3C_3P_2)\}_2Cu]^+$, 326.1 $[Cp^*_2Fe]^+$. |
| **Elemental analysis** | calcd. for $C_{24}H_{39}P_2FeCu_2Br_2$ (746.27 g mol$^{-1}$) C 38.63, H 5.27, N 1.88; found C 39.42, H 5.42, N 2.48. |

### 4.2.6 $^1_\infty[\{Cp^*Fe(\eta^{1:1:5}\text{-}^iPr_3C_3P_2)\}Cu_2(\mu\text{-}I)_2(CH_3CN)_{0.5}]$ (8)

A solution of 32 mg (0.17 mmol) CuI in 6 mL acetonitrile is layered over a solution of 35 mg (0.08 mmol) $[Cp^*Fe(\eta^5\text{-}^iPr_3C_3P_2)]$ (**5**) in 3 mL toluene in a thin schlenk tube. After diffusion, small orange platelets of $^1_\infty[\{Cp^*Fe(\eta^{1:1:5}\text{-}^iPr_3C_3P_2)\}Cu_2(\mu\text{-}I)_2(CH_3CN)_{0.5}]$ (**8**) are formed. The mother liquor is decanted and the crystals are washed with $CH_2Cl_2$ to remove precipitated CuI, then dried under vacuum.

Analytical data for **8**:

| | |
|---|---|
| **Yield** | 7 mg (8.8 µmol, 10 %) |
| **ESI-MS** (toluene/$CH_3CN$) | m/z [%] = 712.0 $[\{Cp^*Fe(^iPr_3C_3P_2)\}Cu_2I(CH_3CN)]^+$, 522.1 $[\{Cp^*Fe(^iPr_3C_3P_2)\}_2Cu(CH_3CN)]^+$, 326.1 $[Cp^*_2Fe]^+$. |
| **Elemental analysis** | calcd. for $C_{22}H_{36}P_2FeCu_2I_2$ (799.21 g mol$^{-1}$) C 33.06, H 4.54; found C 33.76, H 4.54. |

### 4.2.7 $[Cp'''_2P_4]$ (10) and $[Cp^{BIG}{}_2P_4]$ (11)

A suspension of 0.5 mmol $MCp^R$ (NaCp''': 128 mg, $NaCp^{BIG}$: 363 mg) in toluene is added to a solution of 0.5 mmol $FeBr_3$ (148 mg) and 0.25 mmol $P_4$ (31 mg) in toluene. Upon stirring overnight, the dark red solution turns brown and a brownish precipitate forms. After removal of the solvent in vacuum, 1 mL benzene-$d_6$ is added and filtered into a NMR-tube to determine the conversion by $^{31}P$ NMR spectroscopy.

Analytical data for **10** and **11**:

| | |
|---|---|
| **Yields** | **10**: 38 %; **11**: 53 % (calculated from $^{31}P\{^1H\}$ NMR ) |

Further characterisation of **10** and **11** was performed by Dr. C. Schwarzmaier and Dr. S. Heinl.

### 4.2.8 [$Cp^{*}_{2}P_4$] (12)

A suspension of 71 mg (0.5 mmol) LiCp* in toluene is added to a solution of 148 mg $FeBr_3$ (0.5 mmol) and 31 mg $P_4$ (0.25 mmol) in toluene. Upon stirring over night the dark red solution turns brown and a brownish precipitate forms. After removal of the solvent in vacuum, 1 mL benzene-$d_6$ is added and filtered into a NMR-tube to determine the conversion by $^{31}P$ NMR spectroscopy. To obtain crystals for X-ray structure analysis, the residues are first extracted with pentane to separate **12** from LiBr and $FeBr_2$. The solvent is removed under reduced pressure and the residues are now extracted with acetonitrile to separate **12** from unreacted $P_4$. The solvent is again removed under reduced pressure and the residues are dissolved in very little $Et_2O$. A few colourless blocks of **12** are obtained from saturated solutions in diethylether.

Analytical data for **12**:

| | |
|---|---|
| **Yield** | 82% (calculated from $^{31}P\{^{1}H\}$ NMR) |
| (isolated crystalline yield) | a few crystals |
| $^{1}$**H NMR** ($C_6D_6$) | δ [ppm] = 1.02 (s, 3H, C$\underline{H}_3$), 1.65 (s, 6H, C$\underline{H}_3$), 1.89 (s, 6H, C$\underline{H}_3$) |
| $^{31}$**P NMR** ($C_6D_6$) | δ [ppm] = -366.6 (t, $^{1}J_{PP}$ = 193 Hz, 2P, $\underline{P}$-P-Cp*), -144.0 (t, $^{1}J_{PP}$ = 193 Hz, 2P, $\underline{P}$-Cp*). |
| **FD MS** (toluene) | m/z [%] = 394.2 $[M]^{+}$, 135.1 $[Cp^{*}]^{+}$. |
| **Elemental analysis** | calcd. for $C_{20}H_{30}P_4$ (394.35 g $mol^{-1}$) C 60.91 H 7.67; found C 60.91 H 7.65. |

### 4.2.9 [$Cp^{iPr4}_{2}P_4$] (13)

A suspension of 128 mg (0.5 mmol) $NaCp^{4iPr}$ in toluene is added to a solution of 148 mg $FeBr_3$ (0.5 mmol) and 31 mg $P_4$ (0.25 mmol) in toluene. Upon stirring over night the dark red solution turns brown and a brownish precipitate forms. After removal of the solvent in vacuum, 1 mL benzene-$d_6$ is added and filtered into a NMR-tube to determine the conversion by $^{31}P$ NMR spectroscopy. To obtain crystals for X-ray structure analysis, the residues are first extracted with pentane to separate **13** from LiBr and $FeBr_2$. The solvent is removed under reduced pressure and the residues are now extracted with acetonitrile separate **13** from unreacted $P_4$. The solvent is again removed under reduced pressure and a few colourless blocks of **13** are obtained from the oily residue.

Analytical data for **13**:

| | |
|---|---|
| **Yield** | 62 % (calculated from $^{31}P\{^1H\}$ NMR) |
| (isolated crystalline yield) | A few crystals |
| $^{31}$**P NMR** ($C_6D_6$) **A** ($A_2MN$) | δ [ppm] =-140.8 (t, 2P, $^1J$ = 179 Hz, $P_A$), -311.5 (*pseudo*-q, 1P, $^1J$ = 180 Hz, $P_M$), -365.1 (*pseudo*-q, 1P, $^1J$ = 176 Hz, $P_N$). |
| $^{31}$**P NMR** ($C_6D_6$) **B** ($A_2M_2$) | δ [ppm] = -134.2 (t, 2P, $^1J_{AM}$ = 180 Hz, $P_A$), -338.2 (t, 2P, $^1J_{AM}$ = 180 Hz, $P_M$). The signals of Isomer **B** are not well resolved, and therefore only a rough merit can be provided. |
| **FD MS** (toluene) | m/z [%] = 590.4 $[M]^+$, 668.4 $[Cp^{4iPr}{}_2Fe(Et_2O)O]^+$, 710.5 $[Cp^{4iPr}{}_2Fe(Et_2O)(CH_3CN)O]^+$. |

### 4.2.10 **$[CpMn\{\mu\text{-}P(SiMe_3)_2\}]_2$ (15)**

A solution of 100 mg (0.55 mmol) $[Cp_2Mn]$ (**14**) in 10 ml toluene is cooled to –78°C and a solution of 170 mg (0.55 mmol) $Li[P(SiMe_3)_2]$ $(thf)_{1.8}$ in 10 ml toluene is added dropwise. The dark orange reaction mixture was warmed to room temperature and stirred overnight. The resulting solution was filtered (diatomaceous earth) to remove precipitated LiCp. The solution is concentrated by taking off solvent under reduced pressure until saturation is reached. Then, the mixture is stored at –28°C to produce orange parallelepipeds of $[CpMn\{\mu\text{-}P(SiMe_3)_2\}]_2$ (**15**).

Analytical data for **15**:

| | |
|---|---|
| **Yield** | 80 mg (0.13 mmol, 49 %) |
| $^1$**H NMR** ($C_6D_6$) | δ [ppm] = 18.01 ($\omega_{1/2}$ = 797.9 Hz, $C_5\underline{H}_5$), 7.2 (weak, broad peak obscured by the solvent resonance), 0.28 (s, $Si\underline{Me}_3$), 0.23 (s,$Si\underline{Me}_3$). |
| **Elemental analysis** | calcd. for $C_{22}H_{46}P_2Si_4Mn_2$ (594.77 g $mol^{-1}$) C 44.43, H 7.80, P 10.42; found C 43.98, H 7.75, P 10.11. |

### 4.2.11 **$[CpMn\{\mu\text{-}As(SiMe_3)_2\}]_2$ (16)**

A solution of 100 mg (0.55 mmol) $[Cp_2Mn]$ (**14**) in 10 ml toluene is cooled to –78°C and a solution of 201 mg (0.55 mmol) $Li[As(SiMe_3)_2]\cdot(thf)_{1.4}$ in 10 ml toluene is added dropwise. The orange reaction mixture was warmed to room temperature and stirred overnight. The resulting solution was filtered (diatomaceous earth) to remove precipitated LiCp. The solution is concentrated by taking off the solvent under reduced pressure until saturation is reached. Then, the mixture is stored at −28°C to produce orange parallelepipeds of $[CpMn\{\mu\text{-}As(SiMe_3)_2\}]_2$ (**16**).

Analytical data for **16**:

| | |
|---|---|
| **Yield** | 70 mg (0.10 mmol, 37 %) |
| **$^1$H NMR** ($C_6D_6$) | δ [ppm] = 24.15 (br, $C_5\underline{H}_5$) 12.07 ppm (br, overlapping, $C_5\underline{H}_5$), 0.29 (s, $Si\underline{Me}_3$). |
| **Elemental analysis** | calcd. for $C_{22}H_{46}As_2Si_4Mn_2$ (682.67 g mol$^{-1}$) C 42.09, H 7.39; found C 41.87, H 7.21. |

## 4.2.12 $^1_\infty[(\mu;\eta^{2:5}\text{-}Cp)(Cr\{\mu\text{-}N(SiMe_3)_2\}_2Li]$ (18)

A solution of 100 mg (0.52 mmol) $[Cp_2Cr]$ in 10 mL toluene is cooled to −78 °C, and a solution of 340 mg (1.10 mmol) $Li[N(SiMe_3)_2]$ in 20 mL toluene is added dropwise. The blue-black reaction mixture is warmed to room temperature and stirred overnight. The resulting very dark blue solution is filtered (diatomaceous earth) to remove precipitated of LiCp. The filtrate is concentrated under reduced pressure until saturated, and then stored at −28 °C to produce dark violet plates of $^1_\infty[(\mu;\eta^{2:5}\text{-}Cp)(Cr\{\mu\text{-}N(SiMe_3)_2\}_2Li]$ (**18**).

Analytical data for **18**:

| | |
|---|---|
| **Yield** | 50 mg (0.11 mmol, 21 %) |
| **$^1$H NMR** ($C_6D_6$) | δ [ppm] = 23.41 (br, $\omega_{1/2}$ = 1073 Hz, $C_5\underline{H}_5$), 2.85 (br, $\omega_{1/2}$ = 149 Hz, $Si\underline{Me}_3$), 1.55 ($\omega_{1/2}$ = 40 Hz, $Si\underline{Me}_3$). |
| **$^7$Li NMR** (pyridine-$d_5$) | δ [ppm] = −5.1. |
| **Elemental analysis** | calcd. for $C_{17}H_{41}N_2Si_4LiCr$ (444.81 g mol$^{-1}$) C 45.90 H 9.29, N 6.30; found C 45.62, H 8.99, N 6.11. |

## 4.2.13 $[CpCr\{\mu\text{-}P(SiMe_3)_2\}]_2$ (19)

A solution of 100 mg (0.52 mmol) $[Cp_2Cr]$ in 10 mL toluene is cooled to −78 °C, and a solution of 186 mg (0.52 mmol) $Li[P(SiMe_3)_2]\cdot(thf)_{1.8}$ in 10 mL toluene is added dropwise. The dark reaction mixture is warmed to room temperature and stirred overnight. The resulting very dark violet solution is filtered (diatomaceous earth) to remove precipitated of LiCp. The filtrate is concentrated under reduced pressure until saturated, and then stored at −28 °C to produce dark red crystals of $[CpCr\{\mu\text{-}P(SiMe_3)_2\}]_2$ (**19**).

Analytical data for **19**:

| | |
|---|---|
| **Yield** | 80 mg (0.14 mmol, 52 %) |
| **$^1$H NMR** ($C_6D_6$) | δ [ppm] = 98.82 (br, low intensity), 3.35 ($\omega_{1/2}$ = 219.5 Hz, $C_5\underline{H}_5$), 0.28 (s, $\omega_{1/2}$ = 5.2 Hz, $Si\underline{Me}_3$) |
| **$^{31}$P NMR** ($C_6D_6$) | No signal could be detected. |
| **Elemental analysis** | calcd. for $C_{22}H_{46}P_2Si_4Cr_2$ (588.89 g $mol^{-1}$) C 44.87, H 7.87, P 10.52; found C 44.23, H 7.83, P 10.36. |

### 4.2.14 $[CpCr\{\mu\text{-}As(SiMe_3)_2\}]_2$ (20)

A solution of 100 mg (0.52 mmol) $[Cp_2Cr]$ in 10 mL toluene is cooled to −78 °C, and a solution of 194 mg (0.52 mmol) $Li[As(SiMe_3)_2]\cdot(thf)_{1.4}$ in 10 mL toluene is added dropwise. The dark reaction mixture is warmed to room temperature and stirred overnight. The resulting very dark violet solution is filtered (diatomaceous earth) to remove precipitated of LiCp. The filtrate is concentrated under reduced pressure until saturated, and then stored at −28 °C to produce dark red crystals of $[CpCr\{\mu\text{-}As(SiMe_3)_2\}]_2$ (**20**).

Analytical data for **20**:

| | |
|---|---|
| **Yield** | 80 mg (0.12 mmol, 45 %) |
| **$^1$H NMR** ($C_6D_6$) | δ [ppm] = 34.02, (br, very low intensity, $\omega_{1/2}$ = 466.7 Hz), 2.38 ($\omega_{1/2}$ = 82.1 Hz, $C_5\underline{H}_5$), 0.28 (s, $\omega_{1/2}$ = 5.3 Hz, $Si\underline{Me}_3$). |
| **Elemental analysis** | calcd. for $C_{22}H_{46}As_2Si_4Cr_2$ (676.78 g $mol^{-1}$) C 39.04, H 6.85; found C 39.20, H 6.72; |

### 4.2.15 $[Cp^*Cr(\mu_3\text{-}P)]_4$ (21)

75 mg (0.5 mmol) LiCp* and 65 mg (0.5 mmol) $CrCl_2$ are combined as solids, cooled to −78°C, and 10 mL thf is added. The mixture is warmed to room temperature, giving a blue solution, which is

stirred for one hour and then cooled to −78°C again. A solution of 158 mg (0.5 mmol) $Li[P(SiMe_3)_2]\cdot(thf)_{0.98}$ in 5 mL thf is added dropwise, and the reaction is warmed to room temperature. After stirring overnight, the solvent is removed from the dark brown reaction mixture and the residue dissolved in toluene. The solution is filtered (diatomaceous earth), the volume of the filtrate is reduced until saturation is reached, and stored at −28 °C to produce dark violet crystals of $[Cp^*Cr(\mu_3\text{-}P)]_4$ (**21**).

Analytical data for **21**:

| | |
|---|---|
| **Yield** | 36 mg (0.05 mmol, 33 %) |
| **$^1$H NMR** ($C_6D_6$) | δ [ppm] = 47.97 (br, low intensity), 38.27 (br, low intensity); unambiguous assignment of the signals is hampered by the paramagnetic nature of **21** (*cf.* appendix) |
| **Melting point** | Decomposition at 156 °C. |
| **Elemental analysis** | calcd. for $C_{40}H_{60}P_4Cr_4$ (872.80 g $mol^{-1}$) C 55.05, H 6.93, P 14.20; found C 54.97, H 6.83, P 13.91. |

### 4.2.16 $[(Cp^*Cr)_3(\mu_3\text{-}As)_2]$ (22)

75 mg (0.5 mmol) LiCp* and 65 mg (0.5 mmol) $CrCl_2$ are combined as solids, cooled to −78°C, and 10 mL thf is added. The mixture is warmed to room temperature, giving a blue solution, which is stirred for one hour and then cooled to −78°C again. A solution of 170 mg (0.5 mmol) $Li[As(SiMe_3)_2]\cdot(thf)_{0.94}$ in 5 mL thf is added dropwise, and the reaction is warmed to room temperature. After stirring overnight, the solvent is removed from the dark brown reaction mixture and the residue dissolved in toluene. The solution is filtered (diatomaceous earth), the volume of the filtrate is reduced until saturation is reached, and stored at −28 °C to produce dark green crystals of $[(Cp^*Cr)_3(\mu_3\text{-}As)_2]$ (**22**).

Analytical data for **22**:

| | |
|---|---|
| **Yield** | 66 mg (0.09 mmol, 56 %) |
| **$^1$H NMR** ($C_6D_6$) | δ [ppm] = 3.56 (br, low intensity); unambiguous assignment of the signals is hampered by the paramagnetic nature of **21** (*cf.* appendix) |

| | |
|---|---|
| **Melting point** | Decomposition at 168 °C. |
| **Elemental analysis** | calcd. for $C_{30}H_{45}As_2Cr_3$ (711.52 g mol-1) C 50.64, H 6.37;<br>found C 50.22, H 6.23. |

### 4.2.17 **[(Cp*Fe)$_3$($\mu_3$-P)$_2$] (23)**

142 mg (1.0 mmol) LiCp* and 127 mg (1.0 mmol) $FeCl_2$ are combined as solids, cooled to −78°C, and 20 mL thf is added. The mixture is warmed to room temperature, giving a green solution, which is stirred for one hour and then cooled to −78°C again. A solution of 299 mg (1.0 mmol) Li[P($SiMe_3$)$_2$]·(thf)$_{0.98}$ in 10 mL thf is added dropwise, and the reaction is warmed to room temperature. After stirring overnight, the solvent is removed from the very dark brown reaction mixture and the residue dissolved in toluene. The solution is filtered (diatomaceous earth,) the volume of the filtrate is reduced until saturation is reached, and stored at −28 °C to produce black crystals of [(Cp*Fe)$_3$($\mu_3$-P)$_2$] (**23**).

Analytical data for **23**:

| | |
|---|---|
| **Yield** | 50 mg (0.07 mmol, 23 %) |

### 4.2.18 **[(Cp*Fe)$_3$($\mu_3$-As)$_2$] (24)**

142 mg (1.0 mmol) LiCp* and 127 mg (1.0 mmol) $FeCl_2$ are combined as solids, cooled to −78°C, and 20 mL thf is added. The mixture is warmed to room temperature, giving a green solution, which is stirred for one hour and then cooled to −78°C again. A solution of 346 mg (1.0 mmol) Li[As($SiMe_3$)$_2$]·(thf)$_{1.4}$ in 10 mL thf is added dropwise, and the reaction is warmed to room temperature. After stirring overnight, the solvent is removed from the very dark brown reaction mixture and the residue dissolved in toluene. The solution is filtered (diatomaceous earth), the volume of the filtrate is reduced until saturation is reached, and stored at −28 °C to produce black crystals of [(Cp*Fe)$_3$($\mu_3$-As)$_2$] (**24**).

Analytical data for **24**:

| | |
|---|---|
| **Yield** | 45 mg (0.06 mmol, 19 %) |

# 5. Crystallographic Section

## 5.1 General remarks

Single crystal X-ray diffraction analyses of the listed compounds were performed by the following persons:

| | |
|---|---|
| 12, 24 | the author |
| 2, 3, 4, 8 | Dr. Alexander V. Virovets and Dr. Eugenia Peresypkina |
| 7, 13, 15 | Dr. M. Bodensteiner |
| 16, 18, 19, 20 and 21 | Dr. C. Schwarzmaier |
| 22 | Dr. C. Grassl |
| 23 | M.Sc. M. Fleischmann |

The crystallographic data was acquired either at an Agilent Technologies (formerly Oxford diffraction) Gemini R Ultra diffractometer using Cu radiation from sealed tubes and a ruby CCD detector or at an Agilent SuperNova device using a microfocus Cu source with an Atlas CCD detector. The Gemini device was equipped with an Oxford diffraction Cryojet cooler and the SuperNova diffractometer with a Cryostream600 cooling system.

Figures of the molecular structures generated from crystallographic data were prepared with the programs DIAMOND 3.0[112] and SCHAKAL-99.[113]

## 5.2 General procedures

### 5.2.1 Sample handling

Most of the processed crystal samples were air and moisture sensitive. Hence, they were handled in mineral oil (Sigma Aldrich, CAS 8042-47-5) or perfluorinated oil (Fomblin, Sigma Aldrich, CAS 69991-67-9). Appropriate crystals were taken to a CryoLoop (Hampton research) on a goniometer head which was directly attached to the goniometer. Thus, the crystal is brought into the cold nitrogen stream of the cooling system which freezes the surrounding oil and fixes the position of the crystal.

### 5.2.2 Data processing

Integration and data reduction of the measured data was performed using the CrysAlisPro software package.[114] Either semi-empirical multi-scan absorption correction from equivalents or analytical absorption correction from crystal faces was applied after the absorption coefficient was determined from the final structure model.

### 5.2.3 Structure solution and refinement

Structure solution was carried out using direct methods (SHELXS-97[115] or SIR-92[116]) or by charge flipping methods (SUPERFLIP[117]). Least-squares refinement on $F_0^2$ was performed using SHELXL-97. All programs were either implemented in WinGX and SXGraph was used for structure representation during the refinement process,[118] or in the $OLEX^2$ software[119]. In some cases several restraints and constraints had to be used: EADP (equal anisotropic displacement parameters (ADP)), EXYZ (equal location parameters), DELU (similar ADP towards bond direction), ISOR (more isotropic ADP), SADI (similar atomic distance), SIMU (similar ADP).

## 5.3 Crystallographic data for the reported structures

### 5.3.1 $C_{10}H_{16}$@[{Cp*Fe($\eta^{1:1:1:1:1:5}$-$P_5$)}$_{12}$(CuCl)$_{20-y}$] (2)

Compound 2 crystallises as brown blocks from a mixture of acetonitrile and $CH_2Cl_2$ ($R\bar{3}$). The refinement could not be finished up to the present. Therefore, only selected parameters are listed below.

| | | |
|---|---|---|
| Device type | Oxford Diffraction Supernova (Atlas) | |
| Temperature /K | 123(2) | |
| Crystal system | Trigonal | |
| Space group | $R\bar{3}$ | |
| Unit cell dimensions | $a$ = 41.6825(6) Å | $\alpha$ = 90 ° |
| | $b$ = 41.6825(6) Å | $\beta$ = 90 ° |
| | $c$ = 50.0108(9) Å | $\gamma$ = 120 ° |
| Volume /Å$^3$ | 75249(2) | |

### 5.3.2 $C_{10}H_{16}$@[{Cp*Fe($\eta^{1:1:1:1:1:5}$-$P_5$)}$_{12}$(CuBr)$_{20-y}$] (3)

Compound **3** crystallises as brown blocks from a mixture of acetonitrile and $CH_2Cl_2$ ($R\bar{3}$) or toluene ($P\bar{1}$), but only crystals from toluene solutions have been analysed with single crystal X-ray diffraction because of their better quality. The refinement could not be finished up to the present. Therefore, only selected parameters are listed below.

| | | |
|---|---|---|
| Device type | Oxford Diffraction Supernova (Atlas) | |
| Temperature /K | 123(2) | |
| Crystal system | Trigonal | |
| Space group | $R\bar{3}$ | |
| Unit cell dimensions | $a$ = 42.1359(4)Å | $\alpha$ = 90 ° |
| | $b$ = 42.1359(4)Å | $\beta$ = 90 ° |
| | $c$ = 51.8637(5) Å | $\gamma$ = 120 ° |
| Volume /Å$^3$ | 79744(1) | |

| | | |
|---|---|---|
| Empirical formula | $C_{160}H_{240}Br_{18.8}Cl_{1.33}Cu_{18.8}Fe_{12}N_{1.33}P_{60}$ | |
| Formula weight $M$/g mol$^{-1}$ | 7454.71 | |
| Device type | Oxford Diffraction Supernova (Atlas) | |
| Temperature /K | 123(2) | |
| Crystal system | Triclinic | |
| Space group | $P\bar{1}$ | |
| Unit cell dimensions | $a$ = 29.5240(8)Å | $\alpha$ = 90.691(3)° |
| | $b$ = 29.548(1) Å | $\beta$ = 90.417(3) ° |
| | $c$ = 29.832(1) Å | $\gamma$ = 90.386(3) ° |
| Volume /Å$^3$ | 26021.3(1) | |

### 5.3.3 $C_6H_4Cl_2@[\{Cp^*Fe(\eta^{1:1:1:1:1:5}-P_5)\}_8(CuI)_{28}(CH_3CN)_{10}]$ (4)

Compound **3** crystallises as red plates from a mixture of acetonitrile and *ortho*-dichlorobenzene. Due to technical problems, the small period of time the crystals are present in the reaction mixture and their small size, no complete data set could be recorded yet.

| | | |
|---|---|---|
| Empirical formula | $Cl_{20.54}C_{252.73}Cu_{56.02}I_{56}Fe_{16}P_{80}N_{19.25}H_{294}$ | |
| Formula weight $M$/g mol$^{-1}$ | 18334.7 | |
| Device type | Oxford Diffraction Supernova (Atlas) | |
| Temperature /K | 123(2) | |
| Crystal system | Triclinic | |
| Space group | $P\overline{1}$ | |
| Unit cell dimensions | $a$ = 29.5240(8) Å | $\alpha$ = 90.691(3) ° |
| | $b$ = 29.5479(11) Å | $\beta$ = 90.417(3) ° |
| | $c$ = 29.8318(13) Å | $\gamma$ = 90.386(3) ° |
| Volume /Å$^3$ | 26021.2(16) | |
| Z | 3 | |
| $\rho_{calc}$ g/cm$^3$ | 1.427 | |
| $\mu$ /mm$^{-1}$ | 10.567 | |

### 5.3.4 $^1_\infty[\{Cp^*Fe(\eta^{1:1:5}-{}^iPr_3C_3P_2)\}Cu_2(\mu\text{-}Cl)_2(CH_3CN)]$ (6)

Compound **6** crystallises as orange platelets from a 1:1-mixture of toluene and acetonitrile. Due to technical problems, only the unit cell could be determined.

| | | |
|---|---|---|
| Device type | Oxford Diffraction Supernova (Atlas) | |
| Temperature /K | 123(2) | |
| Crystal system | Monoclinic | |
| Space group | - | |
| Unit cell dimensions | $a$ = 15.744 Å | $\alpha$ = 90 ° |
| | $b$ = 19.095 Å | $\beta$ = 92.58 ° |
| | $c$ = 20.306 Å | $\gamma$ = 90 ° |
| Volume /Å$^3$ | 6099 | |

### 5.3.5 $^{1}_{\infty}[\{Cp^{*}Fe(\eta^{1:1:5}\text{-}^{i}Pr_{3}C_{3}P_{2})\}Cu_{2}(\mu\text{-}Br)_{2}(CH_{3}CN)]$ (7)

Compound **7** crystallises as orange platelets from a 1:1-mixture of toluene and acetonitrile.

| | | |
|---|---|---|
| Empirical formula | $C_{25.5}H_{41.25}Br_2Cu_2FeN_{1.75}P_2$ | |
| Formula weight | 777.04 | |
| Device type | Oxford Diffraction Gemini Ultra (Atlas) | |
| Temperature /K | 123.0(10) | |
| Crystal system | Monoclinic | |
| Space group | C2/c | |
| Unit cell dimensions | $a$ = 25.4094(2) Å | $\alpha$ = 90 ° |
| | $b$ = 19.3047(2) Å | $\beta$ = 126.782(1) ° |
| | $c$ = 15.7887(1) Å | $\gamma$ = 90 ° |
| Volume /Å$^3$ | 6202.87(12) | |
| Z | 8 | |
| $\rho_{calc}$ g/cm$^3$ | 1.664 | |
| $\mu$ /mm$^{-1}$ | 9.405 | |
| F(000) | 3124.0 | |
| Crystal size /mm$^3$ | 0.1729 × 0.0584 × 0.05 | |
| Radiation λ/Å | Cu $K_\alpha$ (λ = 1.54178) | |
| Theta range for data collection /° | 7.24 to 133.94 | |
| Index ranges | $-30 \le h \le 30$, $-22 \le k \le 22$, $-18 \le l \le 18$ | |
| Reflections collected | 49865 | |
| Independent reflections | 5508 [$R_{int}$ = 0.0399] | |
| Data/restraints/parameters | 5508/18/345 | |
| Goodness-of-fit on F$^2$ | 1.061 | |
| Final R indexes [I>=2σ (I)] | $R_1$ = 0.0299, $wR_2$ = 0.0741 | |
| Final R indexes [all data] | $R_1$ = 0.0344, $wR_2$ = 0.0768 | |
| Largest diff. peak/hole / e Å$^{-3}$ | 0.97/-0.46 | |

### 5.3.6 $^{1}_{\infty}[\{Cp^*Fe(\eta^{1:1:5}\text{-}{}^iPr_3C_3P_2)\}Cu_2(\mu\text{-}I)_2(CH_3CN)_{0.5}]$ (8)

Compound **8** crystallises as orange platelets from a 3:5-mixture of toluene and acetonitrile. Twin refinement for inversion twinning (BASF 0.594(3), 0.406(3)) was necessary.

| | | |
|---|---|---|
| Empirical formula | $C_{48}H_{77}Cu_4Fe_2I_4N_2P_4$ | |
| Formula weight | 1679.50 | |
| Device type | Oxford Diffraction Gemini Ultra (Atlas) | |
| Temperature /K | 123.0(10) | |
| Crystal system | Orthorhombic | |
| Space group | $P2_12_12_1$ | |
| Unit cell dimensions | $a$ = 11.1640(2) Å | $\alpha$ = 90 ° |
| | $b$ = 14.2251(1) Å | $\beta$ = 90 ° |
| | $c$ = 37.2747(4) Å | $\gamma$ = 90 ° |
| Volume /Å$^3$ | 5919.56(13) | |
| Z | 4 | |
| $\rho_{calc}$ g/cm$^3$ | 1.885 | |
| $\mu$ /mm$^{-1}$ | 23.036 | |
| F(000) | 3276.0 | |
| Crystal size /mm$^3$ | 0.1883 × 0.1321 × 0.042 | |
| Radiation λ/Å | Cu $K_\alpha$ (λ = 1.54184) | |
| Theta range for data collection /° | 6.66 to 134.04 | |
| Index ranges | $-12 \le h \le 13$, $-16 \le k \le 16$, $-44 \le l \le 44$ | |
| Reflections collected | 97318 | |
| Independent reflections | 10518 [$R_{int}$ = 0.0381] | |
| Data/restraints/parameters | 10518/1/602 | |
| Goodness-of-fit on $F^2$ | 1.050 | |
| Final R indexes [I>=2σ (I)] | $R_1$ = 0.0192, $wR_2$ = 0.0456 | |
| Final R indexes [all data] | $R_1$ = 0.0208, $wR_2$ = 0.0459 | |
| Largest diff. peak/hole / e Å$^{-3}$ | 0.85/-0.68 | |

#### 5.3.7 [Cp*$_2$P$_4$] (12)

Compound **12** crystallises from saturated $Et_2O$ solutions as colourless blocks.

| | | |
|---|---|---|
| Empirical formula | $C_{20}H_{30}P_4$ | |
| Formula weight $M$/g mol$^{-1}$ | 394.32 | |
| Device type | Oxford Diffraction Gemini R Ultra | |
| Temperature /K | 123.05(10) | |
| Crystal system | Monoclinic | |
| Space group | $P2_1/c$ | |
| Unit cell dimensions | $a$ = 14.4915(1) Å | $\alpha$ = 90 ° |
| | $b$ = 15.1964(1) Å | $\beta$ = 98.086(1) ° |
| | $c$ = 9.7826(1) Å | $\gamma$ = 90 ° |
| Volume /Å$^3$ | 2132.89(3) | |
| Z | 4 | |
| $\rho_{calc}$ g/cm$^3$ | 1.228 | |
| μ /mm$^{-1}$ | 3.251 | |
| F(000) | 840.0 | |
| Crystal size/mm$^3$ | 0.3632 × 0.2245 × 0.2013 | |
| Radiation λ/Å | Cu $K_\alpha$ (λ = 1.54178) | |
| Theta range for data collection/° | 6.16 to 133.24 | |
| Index ranges | $-17 \le h \le 17$, $-18 \le k \le 17$, $-11 \le l \le 11$ | |
| Reflections collected | 25054 | |
| Independent reflections | 3758 [$R_{int}$ = 0.0280] | |
| Data/restraints/parameters | 3758/0/227 | |
| Goodness-of-fit on $F^2$ | 1.049 | |
| Final R indexes [I>=2σ (I)] | $R_1$ = 0.0264, $wR_2$ = 0.0721 | |
| Final R indexes [all data] | $R_1$ = 0.0282, $wR_2$ = 0.0731 | |
| Largest diff. peak/hole / e Å$^{-3}$ | 0.21/-0.35 | |

### 5.3.8 [$Cp^{iPr4}{}_2P_4$] (13)

Compound **13** crystallises from saturated $Et_2O$ solutions as colourless blocks. Hydrogen atoms were located in idealized positions and refined isotropically according to the riding model. Several tested crystals of **13** were non-merohedrally twinned. Only the major component's reflections (see appendix) were used for the refinement since the second component was much weaker. Hence, HKLF5 refinement results in a much worse model. Therefore the completeness is only 47.8%.

| | | |
|---|---|---|
| Empirical formula | $C_{34}H_{58}P_4$ | |
| Formula weight $M$/g mol$^{-1}$ | 590.68 | |
| Device type | Oxford Diffraction Supernova (Atlas) | |
| Temperature /K | 123.00(10) | |
| Crystal system | Triclinic | |
| Space group | $P\overline{1}$ | |
| Unit cell dimensions | $a$ = 13.37112(18) Å | $\alpha$ = 99.6762(13) ° |
| | $b$ = 17.3342(3) Å | $\beta$ = 103.8043(11) ° |
| | $c$ = 17.7020(2) Å | $\gamma$ = 108.9227(14) ° |
| Volume /Å$^3$ | 3632.05(10) | |
| Z | 4 | |
| $\rho_{calc}$ g/cm$^3$ | 1.080 | |
| $\mu$ /mm$^{-1}$ | 2.050 | |
| F(000) | 1288.0 | |
| Crystal size /mm$^3$ | 0.2585 × 0.1227 × 0.0954 | |
| Radiation λ/Å | Cu $K_\alpha$ (λ = 1.54184) | |
| Theta range for data collection /° | 6.618 to 147.342 | |
| Index ranges | -16 ≤ h ≤ 16, -21 ≤ k ≤ 21, -21 ≤ l ≤ 21 | |
| Reflections collected | 19507 | |
| Independent reflections | 7013 [$R_{int}$ = 0.0429] | |
| Data/restraints/parameters | 7013/12/925 | |
| Goodness-of-fit on F$^2$ | 0.983 | |
| Final R indexes [I>=2σ (I)] | $R_1$ = 0.0382, w$R_2$ = 0.0963 | |
| Final R indexes [all data] | $R_1$ = 0.0501, w$R_2$ = 0.1014 | |
| Largest diff. peak/hole / e Å$^{-3}$ | 0.20/-0.19 | |

### 5.3.9 [CpMn{μ-P($SiMe_3$)$_2$}]$_2$ (15)

Compound **15** crystallises as orange parallelepipeds from saturated toluene solutions upon cooling. The disordered $SiMe_3$ and Cp groups were refined employing SAME, SIMU, DELU and ISOR restraints.

| | | |
|---|---|---|
| Empirical formula | $C_{22}H_{46}Mn_2P_2Si_4$ | |
| Formula weight *M*/g mol$^{-1}$ | 594.77 | |
| Device type | Oxford Diffraction Supernova (Atlas) | |
| Temperature /K | 243(1) | |
| Crystal system | Triclinic | |
| Space group | $P\bar{1}$ | |
| Unit cell dimensions | $a$ = 9.3377(1) Å | $\alpha$ = 90.250(1) ° |
| | $b$ = 11.1617(2) Å | $\beta$ = 90.267(1) ° |
| | $c$ = 17.2118(2) Å | $\gamma$ = 107.955(1) ° |
| Volume /Å$^3$ | 1706.48(4) | |
| Z | 2 | |
| $\rho_{calc}$ g/cm$^3$ | 1.158 | |
| μ /mm$^{-1}$ | 8.303 | |
| F(000) | 628 | |
| Crystal size /mm$^3$ | 0.17 x 0.07 x 0.05 | |
| Radiation λ/Å | Cu $K_\alpha$ (λ = 1.54184) | |
| Theta range for data collection /° | 4.16 to 76.59 | |
| Index ranges | $-11 \leq h \leq 11$, $-12 \leq k \leq 13$, $-21 \leq l \leq 21$ | |
| Reflections collected | 27120 | |
| Independent reflections | 6974 [$R_{int}$ = 0.0412] | |
| Completeness % | 97.2 | |
| Data/restraints/parameters | 6974 / 364 / 493 | |
| Goodness-of-fit on $F^2$ | 1.061 | |
| Final R indexes [I>=2σ (I)] | $R_1$ = 0.0348, $wR_2$ = 0.0954 | |
| Final R indexes [all data] | $R_1$ = 0.0374, $wR_2$ = 0.0983 | |
| Largest diff. peak/hole / e Å$^{-3}$ | 0.292 and -0.390 | |

### 5.3.10 [CpMn{µ-As($SiMe_3$)$_2$}]$_2$ (16)

Compound **16** crystallises as orange parallelepipeds from saturated toluene solutions upon cooling. The disordered $SiMe_3$ and Cp groups were refined employing SAME, SIMU, DELU and ISOR restraints.

| | | |
|---|---|---|
| Empirical formula | $C_{22}H_{46}As_2Mn_2Si_4$ | |
| Formula weight $M$/g mol$^{-1}$ | 682.67 | |
| Device type | Oxford Diffraction Gemini R Ultra | |
| Temperature /K | 123.0(1) | |
| Crystal system | Monoclinic | |
| Space group | $P2_1/n$ | |
| Unit cell dimensions | $a$ = 9.4339(5) Å | $\alpha$ = 90 ° |
| | $b$ = 17.4810(7) Å | $\beta$ = 108.566(5) ° |
| | $c$ = 11.2554(5) Å | $\gamma$ = 90 ° |
| Volume /Å$^3$ | 1759.57(15) | |
| Z | 2 | |
| $\rho_{calc}$ g/cm$^3$ | 1.288 | |
| µ /mm$^{-1}$ | 9.283 | |
| F(000) | 700.0 | |
| Crystal size /mm$^3$ | 0.32 x 0.26 x 0.15 | |
| Radiation λ/Å | Cu $K_\alpha$ (λ = 1.54184) | |
| Theta range for data collection /° | 4.86 to 73.11 | |
| Index ranges | $-11 \le h \le 8$, $-19 \le k \le 21$, $-13 \le l \le 13$ | |
| Reflections collected | 6522 | |
| Independent reflections | 3391 [$R_{int}$ = 0.0346] | |
| Completeness % | 96.4 | |
| Data/restraints/parameters | 3391 / 40 / 187 | |
| Goodness-of-fit on F$^2$ | 1.065 | |
| Final R indexes [I>=2σ (I)] | $R_1$ = 0.0473, $wR_2$ = 0.1274 | |
| Final R indexes [all data] | $R_1$ = 0.0524, $wR_2$ = 0.1335 | |
| Largest diff. peak/hole / e Å$^{-3}$ | 0.797 and -0.582 | |

### 5.3.11 $^1_\infty[(\mu;\eta^{2:5}\text{-Cp})(Cr\{\mu\text{-}N(SiMe_3)_2\}_2Li]$ (18)

Compound **18** crystallises as dark purple plates from saturated toluene solutions upon cooling. The Cp ligands are disordered over two positions, therefore EADP restraints are used for refinement.

| | | |
|---|---|---|
| Empirical formula | $C_{17}H_{41}CrLiN_2Si_4$ | |
| Formula weight $M$/g mol$^{-1}$ | 444.82 | |
| Device type | Oxford Diffraction Gemini R Ultra | |
| Temperature /K | 123.0(10) | |
| Crystal system | Monoclinic | |
| Space group | $P2_1/n$ | |
| Unit cell dimensions | $a$ = 16.7360(2) Å | $\alpha$ = 90 ° |
| | $b$ = 14.8988(2) Å | $\beta$ = 106.434(2) ° |
| | $c$ = 21.2558(3) Å | $\gamma$ = 90 ° |
| Volume /Å$^3$ | 5083.53(13) | |
| Z | 8 | |
| $\rho_{calc}$ g/cm$^3$ | 1.162 | |
| $\mu$ /mm$^{-1}$ | 5.522 | |
| F(000) | 1920.0 | |
| Crystal size /mm$^3$ | 0.3175 × 0.1942 × 0.0724 | |
| Radiation λ/Å | Cu $K_\alpha$ (λ = 1.54178) | |
| Theta range for data collection /° | 5.92 to 133.2 | |
| Index ranges | $-14 \leq h \leq 19$, $-17 \leq k \leq 17$, $-23 \leq l \leq 25$ | |
| Reflections collected | 28031 | |
| Independent reflections | 8792 [$R_{int}$ = 0.0344] | |
| Data/restraints/parameters | 8792/0/442 | |
| Goodness-of-fit on $F^2$ | 0.919 | |
| Final R indexes [I>=2σ (I)] | $R_1$ = 0.0400, $wR_2$ = 0.1053 | |
| Final R indexes [all data] | $R_1$ = 0.0448, $wR_2$ = 0.1074 | |
| Largest diff. peak/hole / e Å$^{-3}$ | 0.70/-0.45 | |

### 5.3.12 [CpCr{μ-P(SiMe$_3$)$_2$}]$_2$ (19)

Compound **19** crystallises as dark red rods from saturated toluene solutions upon cooling.

| | | |
|---|---|---|
| Empirical formula | $C_{22}H_{46}Cr_2P_2Si_4$ | |
| Formula weight $M$/g mol$^{-1}$ | 588.89 | |
| Device type | Oxford Diffraction Supernova (Atlas) | |
| Temperature /K | 123.0(10) | |
| Crystal system | Triclinic | |
| Space group | $P\overline{1}$ | |
| Unit cell dimensions | $a$ = 9.0064(10)Å | $\alpha$ = 90.468(9) ° |
| | $b$ = 10.8050(13) Å | $\beta$ = 90.944(9) ° |
| | $c$ = 17.1383(18) Å | $\gamma$ = 105.076(10) ° |
| Volume /Å$^3$ | 1610.0(3) | |
| Z | 2 | |
| $\rho_{calc}$ g/cm$^3$ | 1.215 | |
| μ /mm$^{-1}$ | 7.981 | |
| F(000) | 624.0 | |
| Crystal size /mm$^3$ | 0.15 × 0.03 × 0.02 | |
| Radiation λ/Å | Cu $K_\alpha$ (λ = 1.54178) | |
| Theta range for data collection /° | 8.48 to 151.3 | |
| Index ranges | $-8 \leq h \leq 11$, $-13 \leq k \leq 12$, $-21 \leq l \leq 17$ | |
| Reflections collected | 11022 | |
| Independent reflections | 6372 [$R_{int}$ = 0.0399] | |
| Data/restraints/parameters | 6372/308/529 | |
| Goodness-of-fit on F$^2$ | 1.009 | |
| Final R indexes [I>=2σ (I)] | $R_1$ = 0.0409, w$R_2$ = 0.0919 | |
| Final R indexes [all data] | $R_1$ = 0.0603, w$R_2$ = 0.1017 | |
| Largest diff. peak/hole / e Å$^{-3}$ | 0.37/-0.52 | |

### 5.3.13 [CpCr{μ-As(SiMe$_3$)$_2$}]$_2$ (20)

Compound **20** crystallises as dark red prisms from saturated toluene solutions upon cooling.

| | | |
|---|---|---|
| Empirical formula | $C_{22}H_{46}As_2Cr_2Si_4$ | |
| Formula weight $M$/g mol$^{-1}$ | 676.79 | |
| Device type | Oxford Diffraction Supernova (Atlas) | |
| Temperature /K | 123.00(10) | |
| Crystal system | Triclinic | |
| Space group | $P\overline{1}$ | |
| Unit cell dimensions | $a$ = 9.1939(3) Å | $\alpha$ = 91.042(3) ° |
| | $b$ = 10.9303(4) Å | $\beta$ = 90.828(3) ° |
| | $c$ = 17.3201(7) Å | $\gamma$ = 106.057(3) ° |
| Volume /Å$^3$ | 1672.01(11) | |
| Z | 2 | |
| $\rho_{calc}$ g/cm$^3$ | 1.344 | |
| μ /mm$^{-1}$ | 8.979 | |
| F(000) | 696.0 | |
| Crystal size /mm$^3$ | 0.1251 × 0.0875 × 0.0509 | |
| Radiation λ/Å | Cu $K_\alpha$ (λ = 1.54184) | |
| Theta range for data collection /° | 8.42 to 152.8 | |
| Index ranges | $-11 \leq h \leq 7$, $-13 \leq k \leq 13$, $-21 \leq l \leq 21$ | |
| Reflections collected | 13531 | |
| Independent reflections | 6764 [$R_{int}$ = 0.0210] | |
| Data/restraints/parameters | 6764/0/273 | |
| Goodness-of-fit on F$^2$ | 1.108 | |
| Final R indexes [I>=2σ (I)] | $R_1$ = 0.0294, $wR_2$ = 0.0806 | |
| Final R indexes [all data] | $R_1$ = 0.0341, $wR_2$ = 0.0838 | |
| Largest diff. peak/hole / e Å$^{-3}$ | 0.65/-0.50 | |

5.3.14 [Cp*Cr($\mu_3$-P)]$_4$ (21)

Compound **21** crystallises as dark violet octahedra from saturated toluene solutions upon cooling. Twin refinement for inversion twinning (BASF 0.494(7), 0.506(7)) and ISOR restraints were necessary.

| | | |
|---|---|---|
| Empirical formula | $C_{40}H_{60}Cr_4P_4$ | |
| Formula weight $M$/g mol$^{-1}$ | 872.76 | |
| Device type | Oxford Diffraction Supernova (Atlas) | |
| Temperature /K | 123.0(10) | |
| Crystal system | Tetragonal | |
| Space group | $I$ | |
| Unit cell dimensions | $a$ = 11.9954(1) Å | $\alpha$ = 90 ° |
| | $b$ = 11.9954(1) Å | $\beta$ = 90 ° |
| | $c$ = 13.9537(3) Å | $\gamma$ = 90 ° |
| Volume /Å$^3$ | 2007.79(5) | |
| Z | 2 | |
| $\rho_{calc}$ g/cm$^3$ | 1.444 | |
| $\mu$ /mm$^{-1}$ | 10.395 | |
| F(000) | 912.0 | |
| Crystal size /mm$^3$ | 0.0717 × 0.0471 × 0.0435 | |
| Radiation λ/Å | Cu $K_\alpha$ (λ = 1.54178) | |
| Theta range for data collection /° | 9.72 to 149 | |
| Index ranges | -14 ≤ h ≤ 14, -14 ≤ k ≤ 14, -16 ≤ l ≤ 16 | |
| Reflections collected | 5961 | |
| Independent reflections | 1985 [$R_{int}$ = 0.0339] | |
| Data/restraints/parameters | 1985/0/115 | |
| Goodness-of-fit on $F^2$ | 1.072 | |
| Final R indexes [I>=2σ (I)] | $R_1$ = 0.0356, $wR_2$ = 0.0922 | |
| Final R indexes [all data] | $R_1$ = 0.0363, $wR_2$ = 0.0929 | |
| Largest diff. peak/hole / e Å$^{-3}$ | 0.47/-0.49 | |

### 5.3.15 [(Cp*Cr)$_3$($\mu_3$-As)$_2$] (22)

Compound **22** crystallises as dark green blocks from saturated toluene solutions upon cooling.

| | | |
|---|---|---|
| Empirical formula | $C_{30}H_{45}As_2Cr_3$ | |
| Formula weight $M$/g mol$^{-1}$ | 711.50 | |
| Device type | Oxford Diffraction Supernova (Atlas) | |
| Temperature /K | 123.0(10) | |
| Crystal system | Orthorhombic | |
| Space group | *Pbca* | |
| Unit cell dimensions | $a$ = 15.3898(3) Å | $\alpha$ = 90 ° |
| | $b$ = 18.4053(3) Å | $\beta$ = 90 ° |
| | $c$ = 21.8004(4) Å | $\gamma$ = 90 ° |
| Volume /Å$^3$ | 6175.05(19) | |
| Z | 8 | |
| $\rho_{calc}$ g/cm$^3$ | 1.531 | |
| $\mu$ /mm$^{-1}$ | 11.039 | |
| F(000) | 2904.0 | |
| Crystal size /mm$^3$ | 0.29 × 0.17 × 0.13 | |
| Radiation λ/Å | Cu $K_\alpha$ (λ = 1.54178) | |
| Theta range for data collection /° | 8.12 to 148.54 | |
| Index ranges | $-19 \le h \le 19$, $-22 \le k \le 22$, $-26 \le l \le 26$ | |
| Reflections collected | 119504 | |
| Independent reflections | 6281 [$R_{int}$ = 0.0419] | |
| Data/restraints/parameters | 6281/0/331 | |
| Goodness-of-fit on $F^2$ | 1.031 | |
| Final R indexes [I>=2σ (I)] | $R_1$ = 0.0280, $wR_2$ = 0.0754 | |
| Final R indexes [all data] | $R_1$ = 0.0284, $wR_2$ = 0.0757 | |
| Largest diff. peak/hole / e Å$^{-3}$ | 0.47/-0.92 | |

### 5.3.16 [(Cp*Fe)$_3$($\mu_3$-P)$_2$] (23)

Compound **23** crystallises as black plates from saturated toluene solutions upon cooling. Twin refinement for inversion twinning was necessary (BASF 0.256(4), 0.744(4)).

| | | |
|---|---|---|
| Empirical formula | $C_{30}H_{45}Fe_3P_2$ | |
| Formula weight $M$/g mol$^{-1}$ | 635.15 | |
| Device type | Oxford Diffraction Supernova (Atlas) | |
| Temperature /K | 123.0(10) | |
| Crystal system | Hexagonal | |
| Space group | $P6_5$ | |
| Unit cell dimensions | $a$ = 10.7059(1) Å | $\alpha$ = 90 ° |
| | $b$ = 10.7059(1) Å | $\beta$ = 90 ° |
| | $c$ = 45.2293(6) Å | $\gamma$ = 120 ° |
| Volume /Å$^3$ | 4489.49(8) | |
| Z | 6 | |
| $\rho_{calc}$ g/cm$^3$ | 1.410 | |
| $\mu$ /mm$^{-1}$ | 12.627 | |
| F(000) | 1998.0 | |
| Crystal size /mm$^3$ | 0.1642 × 0.0886 × 0.0348 | |
| Radiation λ/Å | Cu K$_\alpha$ (λ = 1.54178) | |
| Theta range for data collection /° | 9.54 to 132.74 | |
| Index ranges | $-12 \le h \le 12$, $-8 \le k \le 10$, $-44 \le l \le 52$ | |
| Reflections collected | 19704 | |
| Independent reflections | 4893 [$R_{int}$ = 0.0243] | |
| Data/restraints/parameters | 4893/1/317 | |
| Goodness-of-fit on F$^2$ | 1.035 | |
| Final R indexes [I>=2σ (I)] | $R_1$ = 0.0293, w$R_2$ = 0.0763 | |
| Final R indexes [all data] | $R_1$ = 0.0300, w$R_2$ = 0.0769 | |
| Largest diff. peak/hole / e Å$^{-3}$ | 0.30/-0.37 | |

### 5.3.17 [(Cp*Fe)$_3$($\mu_3$-As)$_2$] (24)

Compound **24** crystallises as brown blocks from saturated toluene solutions upon cooling. Twin refinement for inversion twinning (BASF 0.410(3), 0.590(3)) and ISOR restraints were necessary.

| | | |
|---|---|---|
| Empirical formula | $C_{30}H_{45}As_2Fe_3$ | |
| Formula weight *M*/g mol$^{-1}$ | 723.05 | |
| Device type | Oxford Diffraction Supernova (Atlas) | |
| Temperature /K | 123.0(10) | |
| Crystal system | Hexagonal | |
| Space group | *P6$_1$* | |
| Unit cell dimensions | *a* = 10.76300 Å | *α* = 90 ° |
| | *b* = 10.76300 Å | *β* = 90 ° |
| | *c* = 45.2149(4) Å | *γ* = 120 ° |
| Volume /Å$^3$ | 4536.06(4) | |
| Z | 6 | |
| $\rho_{calc}$ g/cm$^3$ | 1.588 | |
| $\mu$ /mm$^{-1}$ | 13.929 | |
| F(000) | 2214.0 | |
| Crystal size /mm$^3$ | 0.1 × 0.09 × 0.07 | |
| Radiation | Cu $K_\alpha$ ($\lambda$ = 1.54178) | |
| Theta range for data collection /° | 9.48 to 148.66 | |
| Index ranges | $-10 \le h \le 11$, $-13 \le k \le 13$, $-47 \le l \le 55$ | |
| Reflections collected | 17957 | |
| Independent reflections | 5799 [$R_{int}$ = 0.0181] | |
| Data/restraints/parameters | 5799/7/317 | |
| Goodness-of-fit on F$^2$ | 1.099 | |
| Final R indexes [I>=2σ (I)] | $R_1$ = 0.0259, w$R_2$ = 0.0701 | |
| Final R indexes [all data] | $R_1$ = 0.0267, w$R_2$ = 0.0703 | |
| Largest diff. peak/hole / e Å$^{-3}$ | 0.46/-0.45 | |

# 6. Conclusions

## Coordination chemistry of $P_n$ ligand complexes

### Spherical supramolecules from [Cp*Fe($\eta^5$-$P_5$)] (1) and copper(I) halides

In the field of supramolecular chemistry, the assembly of spherical aggregates built up from [Cp*Fe($\eta^5$-$P_5$)] (**1**) and copper(I) halides was investigated: Firstly in the presence of adamantane to see whether an enclosure is possible and secondly to get deeper knowledge about the aggregation processes that take place during the merging of the reactant solutions.

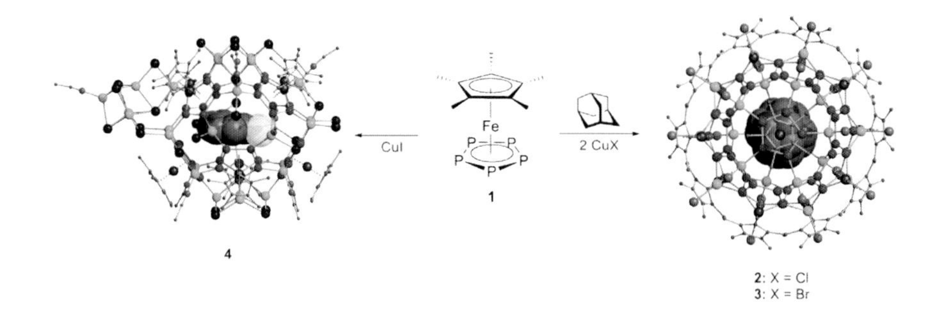


The reaction of pentaphosphaferrocene **1** with two equivalents of CuCl or CuBr in the presence of adamantane leads to the spherical supramolecule $C_{10}H_{16}$@[{Cp*Fe($\eta^5$-$P_5$)}$_{12}$(CuCl)$_{20-y}$] (**2**) or $C_{10}H_{16}$@[{Cp*Fe($\eta^5$-$P_5$)}$_{12}$(CuBr)$_{20-y}$] (**3**), respectively. The $^1$H MAS NMR spectra of the products reveal an upfield shift of the resonance signal of the adamantane protons resulting either from a transfer of electron density to the guest molecule or a shielding effect of the enclosing scaffold. As the cavity of the supramolecule does not fit the cycloalkane well in size, allows it to reorientate without predominant orientation and the framework is also observed with very different guest molecules, very weak host/guest interactions seem to be sufficient for the formation of the '80 vertex ball'.

### Coordination polymers from [Cp*Fe($\eta^5$-$^i$Pr$_3$C$_3$P$_2$)] (5) and copper(I) halides

The coordination behaviour of 1,3-diphosphaferrocenes has not aroused as much interest as that of its relative 1,2,4-triphosphaferrocene, though both derivatives are often obtained in the same synthesis. Hence, the reactivity of [Cp*Fe($\eta^5$-$^i$Pr$_3$C$_3$P$_2$)] (**5**) towards copper(I) halides was studied in this thesis.

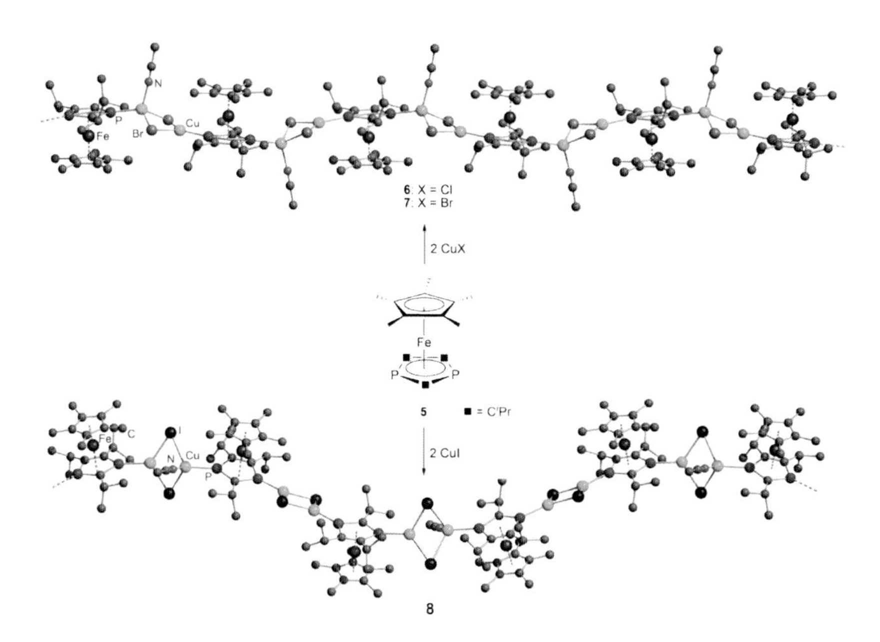


The diphosphaferrocene **5** forms one-dimensional polymers, which all show the well known connectivity pattern with $Cu_2X_2$ four-membered rings as a bridging unit, when reacted with two equivalents of CuX (X = Cl, Br,I). The products **6**, **7**, and **8** differ in their amount of coordinated acetonitrile ligands resulting in different undulations of the polymeric stands in $^1_\infty$[(Cp*Fe($\eta^5$-$^i$Pr$_3$C$_3$P$_2$)Cu$_2$($\mu$-X)$_2$(CH$_3$CN)] (**6**: X = Cl, **7**: X = Br) and $^1_\infty$[(Cp*Fe($\eta^5$-$^i$Pr$_3$C$_3$P$_2$)Cu$_2$($\mu$-X)$_2$(CH$_3$CN)$_{0.5}$] (**8**).

### Iron-mediated C-P bond formation starting from $P_4$

Though the fragmentation and reaggregation of white phosphorus has been widely studied and even though the formation of P–C bonds with $sp^2$-hybridised carbon atoms has been realised, no selective method for the activation of $P_4$ involving $sp^3$ carbon atoms has been known so far. Two observations in our group set the starting point to inquire the activity of iron(III) towards the selective building of P–C bonds: $[Cp'''Fe^{II}(\mu\text{-}Br)]_2$ (**9a**) seemed to disporportionate in the presence of white phosphorus into the active species $[Cp'''Fe^{III}Br_2]$ (and $[Cp'''Fe^{I}]$) causing the formation of $[Cp'''_2P_4]$ (**10**) probably via a radical mechanism. Furthermore, $\{Cp^{BIG}\}^\bullet$ radicals–generated with CuBr and $NaCp^{BIG}$–have been found to attack one of the P–P bonds to form $[Cp^{BIG}_2P_4]$ (**11**).

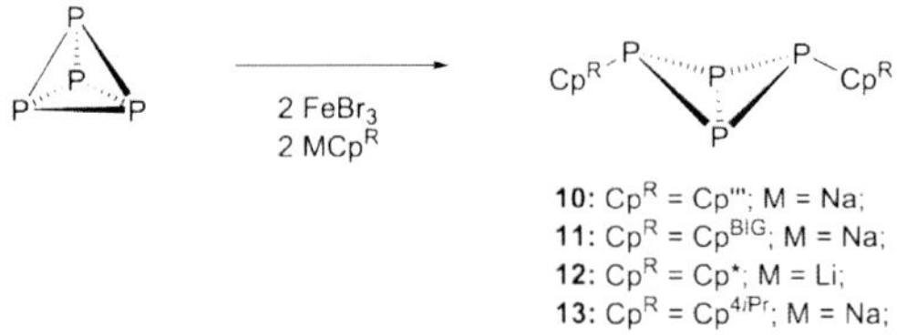

P P P P

2 $FeBr_3$
2 $MCp^R$

$Cp^R$ P P P P $Cp^R$

**10**: $Cp^R$ = Cp'''; M = Na;
**11**: $Cp^R$ = $Cp^{BIG}$; M = Na;
**12**: $Cp^R$ = Cp*; M = Li;
**13**: $Cp^R$ = $Cp^{4iPr}$; M = Na;

P C

**12**

P C

**13**

Thus, during the work for the present thesis, white phosphorus was reacted at room temperature with in situ prepared $[Cp^RFe^{III}Br_2]$ resulting selectively in the already known tetraphospha-bicyclo-butane derivatives **10** and **11**. In addition, this radical generation and transfer in the coordination sphere of the iron(III) centre enables the synthesis of the new 'butterfly' molecules $[Cp^*_2P_4]$ (**12**) and $[Cp^{4iPr}_2P_4]$ (**13**). The further characterisation of **13** revealed as well, that the phosphorus species synthesised by Scherer *et al.* from $P_4$ and $[Cp^{4iPr}CuCO]$ is more likely $[Cp^{4iPr}_2P_4]$ (**13**) than the reported copper complexes $[Cp^{4iPr}Cu(\eta^2\text{-}P_4)]$ and $[Cp^{4iPr}Cu(\mu,\eta^{2:1}\text{-}P_4)CuCp^{4iPr}]$.

By means of this iron-mediated radical transfer, organo-phosphorus compounds can be gained selectively starting from white phosphorus.

### Magnetically active compounds with P- and As-based ligands

New compounds with interesting magnetic properties on a molecular level are in great demand these days. The large majority of the investigated complexes bears oxygen- or nitrogen-based ligands. But because the heavier pnictogens have promising characteristics to mediate magnetic exchange, which have not been well inquired yet, the influence of phosphorus and arsenic as bridging donor atoms was explored in the last part of this thesis.

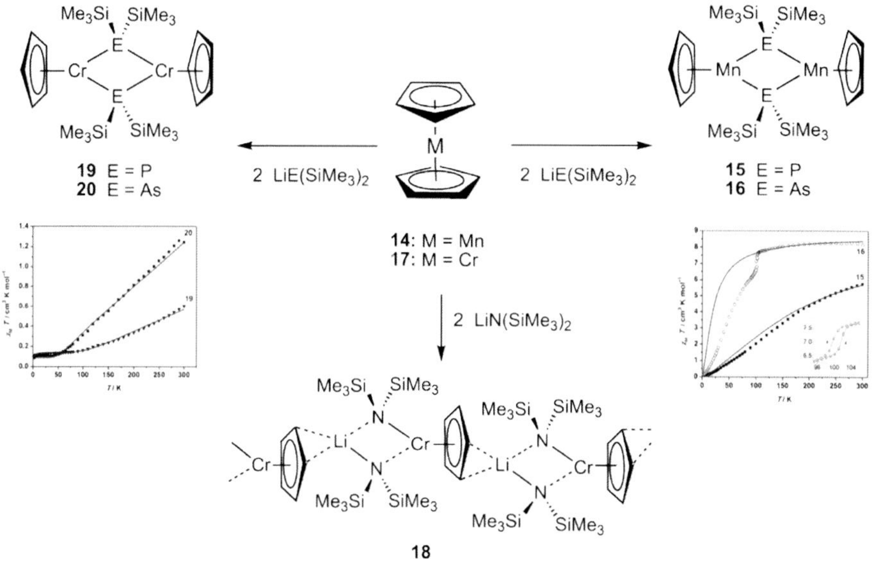

The metallocenes $[Cp_2Mn]$ (**14**) and $[Cp_2Cr]$ (**17**) substitute one of their cyclopentadienyl ligands for a $[E(SiMe_3)_2]^-$ ligand when reacted with $Li[E(SiMe_3)_2]$ (E = N, P, As) to give the binuclear complexes $[CpM\{\mu\text{-}E(SiMe_3)_2\}]_2$ (**15**: M = Mn, E = P; **16**: M = Mn, E = As; **19**: M = Cr; E = P; **20** M = Cr; E = As) and the one-dimensional polymer $^{1}_{\infty}[(\mu{:}\eta^2{:}\eta^5\text{-}Cp)Cr\{\mu\text{-}N(SiMe_3)_2\}_2Li]$ (**18**) , respectively. SQUID measurements reveal that both manganese compounds bear five unpaired electrons that are antiferromagnetically coupled. On the one hand, the coupling constant of the phosphorus derivative **15** is with $J = -13.5\ cm^{-1}$ about one order of magnitude higher than in the arsenic complex **16** where $J = -1.5\ cm^{-1}$. But **16** shows a two step spin crossover including hysteresis behaviour to an intermediate spin state of $S = 3/2$, that probably occurs as well in **15**, only masked by a stronger antiferromagnetic coupling. The investigation of the magnetic behaviour of the chromium compounds displays that all metal centres bear four unpaired electrons and that the $Cr^{II}$ ions are isolated within compound **18** whereas they are strongly antiferromagnetically coupled in **19** ($J = -166\ cm^{-1}$) and **20** ($J = -77.5\ cm^{-1}$). The complexes **16** and **19** most of all confirm the beneficial influence the heavier pnictogens can have as bridging ligands on the magnetic properties: very strong exchange coupling can be mediated by phosphorus as the donor atom, the weaker magnetic exchange via the arsenic ligand enables a spin crossover process.

The introduction of Cp* ligands to induce thermal spin crossover by a stronger ligand field has not been successfully using the respective metallocenes as starting materials and made, therefore, a different approach necessary. The dimers $[Cp^*M(\mu\text{-}Cl)]_2$ (M = Cr, Fe), generated in situ from $MCl_2$ and LiCp*, react with $Li[E(SiMe_3)_2]$ (E = P, As) to the heteroelement cluster compounds $[Cp^*Cr(\mu_3\text{-}P)]_4$ (**21**) and $[(Cp^*M)_3(\mu_3\text{-}E)_2]$ (**22**: M = Cr, E = As; **23**: M = Fe, E = P; **24**: M = Fe, E = As), respectively.

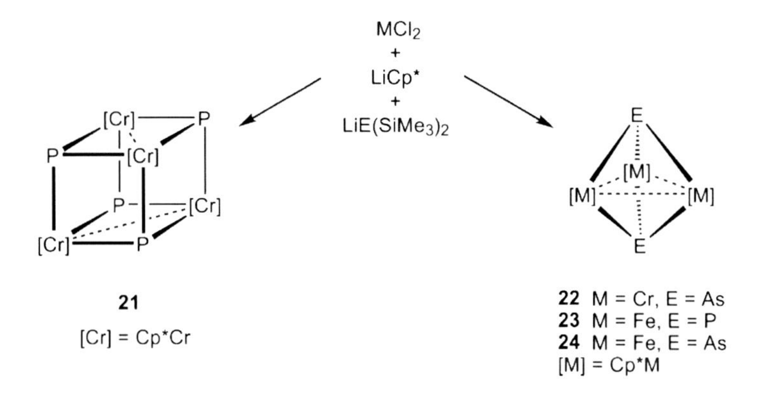


Measurements of the magnetic susceptibility, the magnetisation and the results of DFT calculations suggest that the chromium centres in **21** are in the oxidation state +IV with a spin ground state of S = 2 and chromium-chromium bonding. Compound **22** contains three $Cr^{III}$ ions with a spin ground state of S = 9/2. The complete characterisation of the iron complexes **23** and **24** could unfortunately not be finished within the scope of this work.

This one-pot synthesis seems to provide an easy way to compounds with metal centres in unusual oxidation states like in **21** as well as to heteroelement clusters with 'naked' $\{\mu_3\text{-}E\}^{3-}$ (E = P, As) ligands.

# 7. Appendices

## 7.1 Supplementary Figures

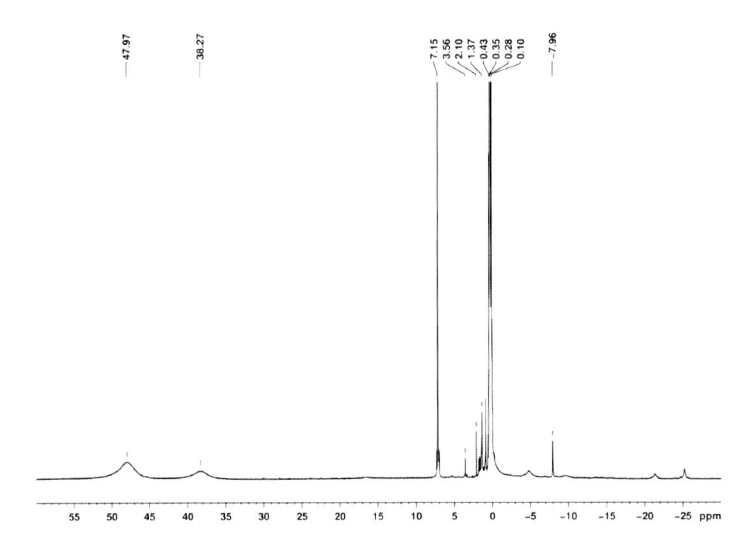


Figure 18: $^{1}H$ NMR spectrum ($C_6D_6$) of 21.

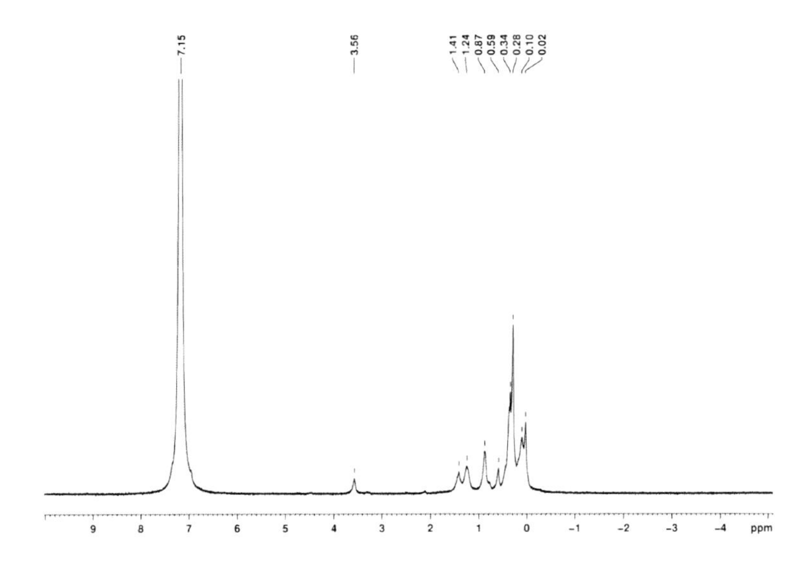


Figure 19: $^{1}H$ NMR spectrum ($C_6D_6$) of 22.

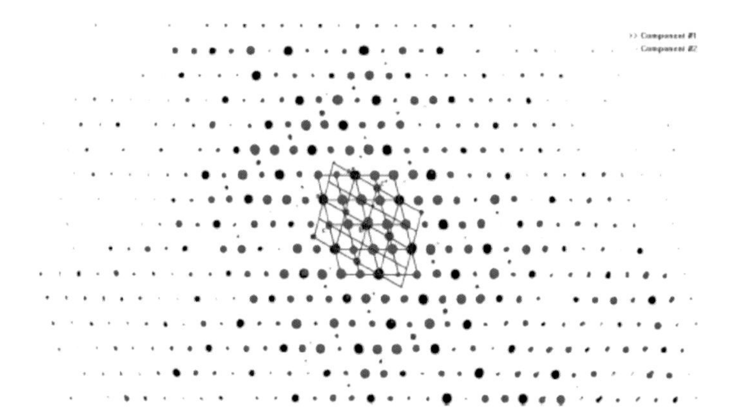


Figure 20: Reciprocal space view along the b* axis (compound **13**). Reflections of component 1 in blue,component 2 in red and overlapping reflections in black.

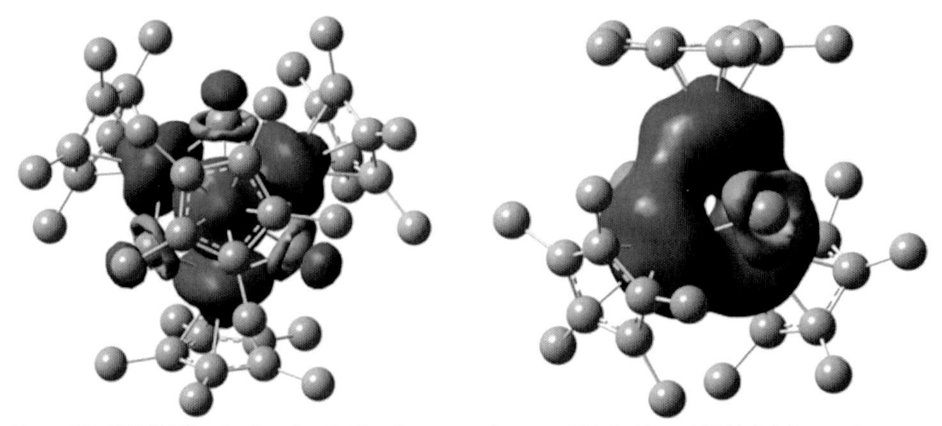

Figure 21: PBE/TZVP spin density plot for the ground state of **21** (left) and **22**(right) (isosurface value = 0.004 a.u.).

## 7.2 List of Abbreviations

| | |
|---|---|
| Å | Angstroem, 1 Å = $1 \cdot 10^{-10}$ m |
| °C | degree Celsius |
| 1D | one dimensional |
| 2D | two dimensional |
| AiM | atoms in molecules |
| $Ar^{Dipp}$ | $-C_6H_3-2,6-(C_6H_3-2,6-{}^iPr_2)_2$ |
| br(NMR) | broad |
| CAAC | *cyclic* (alkyl)-(amino)carbene |
| COSY | correlation spectroscopy |
| Cp | cyclopentadienyl |
| Cp''' | 1,2,4-tris-*tert*-butylcyclopentadienyl |
| Cp* | pentametyhlcyclopentadienyl |
| $Cp^{iPr}$ | penta-isopropylcyclopentadienyl |
| $Cp^{4iPr}$ | tetra-isopropylcyclopentadienyl |
| d(NMR) | doublet |
| $\delta$ | chemical shift |
| DFT | density functional theory |
| Dipp | 2,6-diisopropylphenyl |
| dme | 1,2-dimethoxyethane |
| dppe | 1,2-bis(diphenylphosphino)ethane |
| dppm | 1,2-bis(diphenylphosphino)methane |
| E | heavier element of the 15$^{th}$ group, E = P, As |
| $e^-$ | electron |
| EI MS | electron impact mass spectrometry |
| ESI MS | electron spray ionization |
| Et | ethyl, $-C_2H_5$ |
| $Et_2O$ | diethylether |
| FD MS | field desorption ionization mass spectrometry |
| *H* | magnetic field strength |
| Hz | Hertz |
| $^iPr$ | *iso*-propyl |
| IR | infrared spectroscopy |
| *J*(NMR, SQUID) | coupling constant |

| | |
|---|---|
| kJ | kilo Joule |
| *M* | Magnetisation |
| M | metal |
| *m/z* | mass to charge ratio |
| MAS | magic angle spinning |
| Me | methyl |
| Mes | mesityl, 2,4,6-trimethylphenyl |
| NHC | N-heterocyclic carbene |
| NMR | nuclear magnetic resonance |
| $\tilde{\nu}$ | frequency/wavenumber |
| $\omega_{1/2}$ | full width at half maximum |
| Ph | phenyl |
| ppm | parts per million |
| q(NMR) | quartett |
| R | organic substituent |
| r.t. | room temperature |
| s(NMR) | singlet |
| S | spin quantum number |
| sept(NMR) | septet |
| sMes | 2,4,6-tri-tertbutylphenyl |
| SQUID | Superconducting Quantum Interference Device |
| *T* | temperature |
| t(NMR) | triplet |
| $^{t}$Bu | *tert*-butyl, -$C(CH_3)_3$ |
| thf | tetrahydrofurane, $C_4H_8O$ |
| TMS | tetramethylsilane, $Si(CH_3)_4$ |
| triphos | 1,1,1-tris-(diphenylphopshinomethyl)ethane) |
| vdW | van der Waals |
| VE | valence electron |
| X | any halide, X = Cl, Br, I |
| $\chi_M$ | magnetic susceptibility |
| Z | any light atom |

## 7.3 List of Reported Compounds

1 $[Cp^{*}Fe(\eta^{5}\text{-}P_5)]$

2 $C_{10}H_{16}@[\{Cp^{*}Fe(\eta^{1:1:1:1:1:5}\text{-}P_5)\}_{12}(CuCl)_{20\text{-}y}]$

3 $C_{10}H_{16}@[\{Cp^{*}Fe(\eta^{1:1:1:1:1:5}\text{-}P_5)\}_{12}(CuBr)_{20\text{-}y}]$

4 $C_6H_4Cl_2@[\{Cp^{*}Fe(\eta^{1:1:1:1:1:5}\text{-}P_5)\}_8(CuI)_{28}(CH_3CN)_{10}]$

5 $[Cp^{*}Fe(\eta^{5}\text{-}{}^{i}Pr_3C_3P_2]$

6 $^{1}_{\infty}[\{Cp^{*}Fe(\eta^{1:1:5}\text{-}{}^{i}Pr_3C_3P_2)\}Cu_2(\mu\text{-}Cl)_2(CH_3CN)]$

7 $^{1}_{\infty}[\{Cp^{*}Fe(\eta^{1:1:5}\text{-}{}^{i}Pr_3C_3P_2)\}Cu_2(\mu\text{-}Br)_2(CH_3CN)]$

8 $^{1}_{\infty}[\{Cp^{*}Fe(\eta^{1:1:5}\text{-}{}^{i}Pr_3C_3P_2)\}Cu_2(\mu\text{-}I)_2(CH_3CN)_{0.5}]$

9a $[Cp'''Fe(\mu\text{-}Br)]_2$

9b $[Cp^{iPr4}Fe(\mu\text{-}Br)]_2$

10 $[Cp'''_2P_4]$

11 $[Cp^{BIG}{}_2P_4]$

12 $[Cp^{*}{}_2P_4]$

13 $[Cp^{iPr4}{}_2P_4]$

14 $[Cp_2Mn]$

15 $[CpMn\{\mu\text{-}P(SiMe_3)_2\}]_2$

16 $[CpMn\{\mu\text{-}As(SiMe_3)_2\}]_2$

17 $[Cp_2Cr]$

18 $^{1}_{\infty}[(\mu;\eta^{2:5}\text{-}Cp)(Cr\{\mu\text{-}N(SiMe_3)_2\}_2Li]$

19 $[CpCr\{\mu\text{-}P(SiMe_3)_2\}]_2$

20 $[CpCr\{\mu\text{-}As(SiMe_3)_2\}]_2$

21 $[Cp^{*}Cr(\mu_3\text{-}P)]_4$

22 $[(Cp^{*}Cr)_3(\mu_3\text{-}As)_2]$

23 $[(Cp^{*}Fe)_3(\mu_3\text{-}P)_2]$

24 $[(Cp^{*}Fe)_3(\mu_3\text{-}As)_2]$

$P_{2d}$ $^{2}_{\infty}[\{Cp^{*}Fe(\eta^{1:1:1:5}\text{-}P_5)\}CuX)$ (X = Cl, Br, I)

$P_{odCB}$ $^{1}_{\infty}[\{Cp^{*}Fe(\eta^{1:1:5}\text{-}P_5)\}CuI)]\cdot 0.5C_6H_4Cl_2$

## 7.4 Acknowledgments

Since the end is near, it is time to express my gratitude:

- Prof. Dr. Manfred Scheer for giving me the opportunity to work in his research group.
- Dr. Richard A. Layfield for the fruitful collaboration on the magnetically active compounds, the good lab atmosphere, and all the advice.
- Dr. Gábor Balázs for answers to almost every question, apple strudel and being our heart and soul.
- Prof. Dr. H. Sitzmann for kindly providing a sample of $NaCp^{4iPr}$.
- Dr. Michael Bodensteiner, Dr. Alexander V. Virovets and Dr. Eugenia Peresypkina for their help with problematic X-ray structures. Dr. Christoph Schwarzmaier, Dr. Sebastian Heinl, Michael Seidl und Claudia Heindl for advice, time and patience concerning the everyday problems.
- Prof. Dr. Werner Kremer for the MAS NMR measurements and advice on where to get new rotor caps.
- Dr. Ilya Shenderovich, Anette Schramm, Georgine Stühler and Fritz Kastner for the measurement of my NMR spectra.
- The staff of the micro analytical laboratory for the elemental analyses, even prompt if needed.
- Wolfgang Söllner and Josef Kirmeier for the mass spectometric analyses.
- All the co-workers of the glass blowing, electronics and mechanics facilities of the University of Regensburg.
- The proofreading group: Gábor, Richard and Maria.
- All former and present members of the JCF Regensburg for an unforgettable time and a lot of fireworks.
- Dr. Sebastian Hein(d)l for the formidable trip to Jerusalem.
- The remaining two members of the Wednesday Yoga Trio–crane pose is near.
- My former and present colleagues in the girls´lab–Dr. Andrea, Dr. Thomsi and Claudi for having a wonderful atmosphere.
- All former and present members of the research group for dutifully eating all my "Einfach-so-Kuchen" and the wonderful time: Manfred Zabel (incorporated), Gábor, Joachim Wachter, Karin, Andrea, Conny, Welschi, Oime, Hannes, Fabü, Miriam, Boudi, Stubi, Biegi, Proni, Patrick, Thomsi, Nazhia, Liese, Schotti, Musch, Walter, Petra, Barbara, Matthias, Bianca, Mia, Eric, Christian (Küken), Felix, Rudi, Christian M., Jens, Oli, Martin,

Michi, Luigi, Reinhard, Andi, Wurzel, Basti, Dani, Barbara, Moni, Matthias R.,Mehdi, Susi, Sebi, Moritz, Fabi, Eva and Claudi.

- My family and friends for their happiness, trust and strength

and Bernd for being the sun in my life.

# 8. Notes and References


[1] F. Krafft, *Angew. Chem., Int. Ed. Engl.* **1969**, *8*, 660-671.

[2] a) A. Simon, H. Borrmann, H. Craubner, *Phosphorous and Sulfur and the Related Elements* **1987**, *30*, 507-510; b) A. Simon, H. Borrmann, J. Horakh, *Chem. Ber./Recl.* **1997**, *130*, 1235-1240.

[3] W. L. Roth, T. W. DeWitt, A. J. Smith, *J. Am. Chem. Soc.* **1947**, *69*, 2881-2885.

[4] R. A. L. Winchester, M. Whitby, M. S. P. Shaffer, *Angew. Chem., Int. Ed.* **2009**, *48*, 3616-3621, S3616/3611-S3616/3626.

[5] M. Ruck, D. Hoppe, B. Wahl, P. Simon, Y. Wang, G. Seifert, *Angew. Chem., Int. Ed.* **2005**, *44*, 7616-7619.

[6] a) H. Thurn, H. Krebs, *Angew. Chem., Int. Ed. Engl.* **1966**, *5*, 1047-1048; b) H. Thurn, H. Krebs, *Acta Crystallogr., Sect. B* **1969**, *25*, 125-135.

[7] a) R. Hultgren, N. S. Gingrich, B. E. Warren, *J. Chem. Phys.* **1935**, *3*, 351-355; b) A. Brown, S. Rundqvist, *Acta Crystallogr.* **1965**, *19*, 684-685.

[8] A. Pfitzner, M. F. Braeu, J. Zweck, G. Brunklaus, H. Eckert, *Angew. Chem., Int. Ed.* **2004**, *43*, 4228-4231.

[9] N. Wiberg, *Lehrbuch der anorganischen Chemie, Vol. 101*, de Gruyter, Berlin, New York, **1995**.

[10] C. Schwarzmaier, M. Sierka, M. Scheer, *Angew. Chem., Int. Ed.* **2013**, *52*, 858-861.

[11] H. Erdmann, U. M. v, *Z. anorg. Ch.* **1902**, *32*, 437-452.

[12] J. Eiduss, R. Kalendarev, A. Rodionov, A. Sazonov, G. Chikvaidze, *physica status solidi (b)* **1996**, *193*, 3-23.

[13] A. Bettendorff, *Justus Liebigs Annalen der Chemie* **1867**, *144*, 110-114.

[14] H. Stohr, *Z. Anorg. Allg. Chem.* **1939**, *242*, 138-144.

[15] a) O. Osters, T. Nilges, F. Bachhuber, F. Pielnhofer, R. Weihrich, M. Schöneich, P. Schmidt, *Angewandte Chemie* **2012**, *124*, 3049-3052; b) O. Osters, T. Nilges, F. Bachhuber, F. Pielnhofer, R. Weihrich, M. Schoeneich, P. Schmidt, *Angew. Chem., Int. Ed.* **2012**, *51*, 2994-2997, S2994/2991-S2994/2913.

[16] H. Krebs, W. Holz, K. H. Worms, *Chem. Ber.* **1957**, *90*, 1031-1037.

[17] A. J. Bradley, *Philos. Mag. (1798-1977)* **1924**, *47*, 657-670.

[18] J. Emsley, *The 13th Element: The Sordid Tale of Murder, Fire, and Phosphorus*, John Wiley & Sons, New York, **2000**.

[19] a) M. Caporali, L. Gonsalvi, A. Rossin, M. Peruzzini, *Chem Rev* **2010**, *110*, 4178-4235; b) B. M. Cossairt, N. A. Piro, C. C. Cummins, *Chem Rev* **2010**, *110*, 4164-4177.

[20] M. Scheer, G. Balazs, A. Seitz, *Chem. Rev. (Washington, DC, U. S.)* **2010**, *110*, 4236-4256.

[21] A. P. Ginsberg, W. E. Lindsell, *J. Amer. Chem. Soc.* **1971**, *93*, 2082-2084.

[22] A. Vizi-Orosz, G. Palyi, L. Marko, *J. Organometal. Chem.* **1973**, *60*, C25-C26.

[23] a) M. Di Vaira, C. A. Ghilardi, S. Midollini, L. Sacconi, *J. Am. Chem. Soc.* **1978**, *100*, 2550-2551; b) M. Di Vaira, S. Midollini, L. Sacconi, *J. Am. Chem. Soc.* **1979**, *101*, 1757-1763; c) P. Dapporto, S. Midollini, L. Sacconi, *Angew. Chem.* **1979**, *91*, 510; d) P. Dapporto, S. Midollini, L. Sacconi, *Angewandte Chemie International Edition in English* **1979**, *18*, 469-469.

[24] F. Dielmann, University Regensburg (Regensburg), **2011**.

[25] M. Scheer, *Dalton Transactions* **2008**, 4372-4386.

[26] J. Schwalb, Ph. D. thesis, Universitiy of Kaiserslautern **1988**.

[27] P. Sekar, S. Umbarkar, M. Scheer, A. Voigt, R. Kirmse, *Eur. J. Inorg. Chem.* **2000**, 2585-2589.

[28] a) M. Di Vaira, M. P. Ehses, M. Peruzzini, P. Stoppioni, *Polyhedron* **1999**, *18*, 2331-2336; b) M. Di Vaira, P. Stoppioni, M. Peruzzini, *J. Chem. Soc., Dalton Trans.* **1990**, 109-113; c) F. Cecconi, C. A. Ghilardi, S. Midollini, A. Orlandini, *J. Chem. Soc., Chem. Commun.* **1982**, 229-230.

[29] O. J. Scherer, T. Brück, *Angewandte Chemie* **1987**, *99*, 59-59.

[30] a) J. Bai, A. V. Virovets, M. Scheer, *Science (Washington, DC, U. S.)* **2003**, *300*, 781-783; b) M. Scheer, J. Bai, B. P. Johnson, R. Merkle, A. V. Virovets, C. E. Anson, *Eur. J. Inorg. Chem.* **2005**, 4023-4026; c) M. Scheer, A. Schindler, J. Bai, B. P. Johnson, R. Merkle, R. Winter, A. V. Virovets, E. V. Peresypkina, V. A. Blatov, M. Sierka, H. Eckert, *Chem. - Eur. J.* **2010**, *16*, 2092-2107, S2092/2091-S2092/2096.

[31] M. Scheer, A. Schindler, R. Merkle, B. P. Johnson, M. Linseis, R. Winter, C. E. Anson, A. V. Virovets, *J. Am. Chem. Soc.* **2007**, *129*, 13386-13387.

[32] M. Scheer, A. Schindler, C. Groeger, A. V. Virovets, E. V. Peresypkina, *Angew. Chem., Int. Ed.* **2009**, *48*, 5046-5049, S5046/5041-S5046/5043.

[33] C. Schwarzmaier, A. Schindler, C. Heindl, S. Scheuermayer, E. V. Peresypkina, A. V. Virovets, M. Neumeier, R. Gschwind, M. Scheer, *Angew. Chem., Int. Ed.* **2013**, *52*, 10896-10899.

[34] C. Schwarzmaier, Ph.D. thesis, University of Regensburg (Regensburg), **2012**.

[35] Given diameters are calculated by adding the van der Waals radii of the respective atoms to the distance of two geometrically opposing atoms (H: 1.2 Å, P 1.8 Å,Cu: 1.4 Å, I: 2.0 Å). Distances between *cyclo*-$P_5$ rings refer to the distance between their midpoints.

[36] A. Schindler, Ph.D. thesis, University of Regensburg (Regensburg), **2010**.

[37] J. Bai, A. V. Virovets, M. Scheer, *Angew. Chem., Int. Ed.* **2002**, *41*, 1737-1740.

[38] A. Schindler, C. Heindl, G. Balazs, C. Groeger, A. V. Virovets, E. V. Peresypkina, M. Scheer, *Chem. - Eur. J.* **2012**, *18*, 829-835, S829/821-S829/817.

[39] C. Heindl, unpublished results.

[40] F. Dielmann, A. Schindler, S. Scheuermayer, J. Bai, R. Merkle, M. Zabel, A. V. Virovets, E. V. Peresypkina, G. Brunklaus, H. Eckert, M. Scheer, *Chem. - Eur. J.* **2012**, *18*, 1168-1179, S1168/1161-S1168/1166.

[41] a) C. Mueller, R. Bartsch, A. Fischer, P. G. Jones, *Polyhedron* **1993**, *12*, 1383-1390; b) R. Bartsch, P. B. Hitchcock, J. F. Nixon, *Journal of Organometallic Chemistry* **1988**, *340*, C37-C39; c) C. Mueller, R. Bartsch, A. Fischer, P. G. Jones, R. Schmutzler, *J. Organomet. Chem.* **1996**, *512*, 141-148; d) C. S. J. Callaghan, P. B. Hitchcock, J. F. Nixon, *J. Organomet. Chem.* **1999**, *584*, 87-93.

[42] a) S. Deng, C. Schwarzmaier, U. Vogel, M. Zabel, J. F. Nixon, M. Scheer, *Eur. J. Inorg. Chem.* **2008**, 4870-4874; b) A. Schindler, G. Balazs, M. Zabel, C. Groeger, R. Kalbitzer, M. Scheer, *C. R. Chim.* **2010**, *13*, 1241-1248; c) S. Deng, C. Schwarzmaier, M. Zabel, J. F. Nixon, M. Bodensteiner, E. V. Peresypkina, G. Balázs, M. Scheer, *European Journal of Inorganic Chemistry* **2011**, *2011*, 2991-3001.

[43] R. Bartsch, P. B. Hitchcock, J. F. Nixon, *J. Chem. Soc., Chem. Commun.* **1987**, 1146-1148.

[44] a) H. H. Karsch, *Synthetic Methods of Organometallic and Inorganic Chemistry, Vol. 3*, Thieme Verlag, Stuttgart, **1996**; b) In collaboration with C. Schwarzmaier and C. Heindl

[45] M. M. Rauhut, A. M. Semsel, *J. Org. Chem.* **1963**, *28*, 471-473.

[46] J. D. Masuda, W. W. Schoeller, B. Donnadieu, G. Bertrand, *Angew. Chem., Int. Ed.* **2007**, *46*, 7052-7055.

[47] J. D. Masuda, W. W. Schoeller, B. Donnadieu, G. Bertrand, *J. Am. Chem. Soc.* **2007**, *129*, 14180-14181.

[48] O. Back, G. Kuchenbeiser, B. Donnadieu, G. Bertrand, *Angew. Chem., Int. Ed.* **2009**, *48*, 5530-5533, S5530/5531-S5530/5539.

[49] D. Holschumacher, T. Bannenberg, K. Ibrom, C. G. Daniliuc, P. G. Jones, M. Tamm, *Dalton Trans.* **2010**, *39*, 10590-10592.

[50] E. Fluck, R. Riedel, H. D. Hausen, G. Heckmann, *Z. Anorg. Allg. Chem.* **1987**, *551*, 85-94.

[51] A. R. Fox, R. J. Wright, E. Rivard, P. P. Power, *Angew. Chem., Int. Ed.* **2005**, *44*, 7729-7733.

[52] B. M. Cossairt, C. C. Cummins, *New J. Chem.* **2010**, *34*, 1533-1536.

[53] H. Sitzmann, R. Boese, *Angewandte Chemie International Edition in English* **1991**, *30*, 971-973.

[54] M. Wallasch, G. Wolmershaeuser, H. Sitzmann, *Angew. Chem., Int. Ed.* **2005**, *44*, 2597-2599.

[55] D. N. Akbayeva, O. J. Scherer, *Z. Anorg. Allg. Chem.* **2001**, *627*, 1429-1430.

[56] a) S. Heinl, S. Reisinger, C. Schwarzmaier, M. Bodensteiner, M. Scheer, *Angewandte Chemie International Edition* **2014**, n/a-n/a; b) S. Heinl, S. Reisinger, C. Schwarzmaier, M. Bodensteiner, M. Scheer, *Angewandte Chemie* **2014**, n/a-n/a.

[57] M. Scheer, S. Gremler, E. Herrmann, M. Dargatz, H. D. Schädler, *Zeitschrift für anorganische und allgemeine Chemie* **1993**, *619*, 1047-1052.

[58] P. Kögerler, *Nachrichten aus der Chemie* **2008**, *56*, 743-746.

[59] O. Sato, J. Tao, Y.-Z. Zhang, *Angew Chem Int Ed Engl* **2007**, *46*, 2152-2187.

[60] B. Bleaney, K. D. Bowers, *Proc. R. Soc. London, Ser. A* **1952**, *214*, 451-465.

[61] a) T. Glaser, *Chem. Commun. (Cambridge, U. K.)* **2011**, *47*, 116-130; b) M. Murrie, *Chem. Soc. Rev.* **2010**, *39*, 1986-1995; c) O. Roubeau, R. Clerac, *Eur. J. Inorg. Chem.* **2008**, 4325-4342; d) R.-J. Wei, Q. Huo, J. Tao, R.-B. Huang, L.-S. Zheng, *Angew. Chem., Int. Ed.* **2011**, *50*, 8940-8943, S8940/8941-S8940/8944; e) B. Schneider, S. Demeshko, S. Dechert, F. Meyer, *Angew. Chem., Int. Ed.* **2010**, *49*, 9274-9277, S9274/9271-S9274/9214; f) D.-Y. Wu, O. Sato, Y. Einaga, C.-Y. Duan, *Angew. Chem., Int. Ed.* **2009**, *48*, 1475-1478, S1475/1471-S1475/1478.

[62] a) L. J. Batchelor, E. Fitzgerald, J. Wolowska, J. J. W. McDouall, E. J. L. McInnes, *Chem. - Eur. J.* **2010**, *16*, 11082-11088, S11082/11081-S11082/11083; b) J. R. Rambo, S. L. Castro, K. Folting, S. A. Bartley, R. Heintz, G. Christou, *Inorg. Chem.* **1996**, *35*, 6844-6852; c) N. S. Dean, S. L. Bartley, W. E. Streib, E. B. Lobkovsky, G. Christou, *Inorg. Chem.* **1995**, *34*, 1608-1616; d) C. A. Smith, F. Tuna, M. Bodensteiner, M. Helliwell, D. Collison, R. A. Layfield, *Dalton Trans.* **2013**, *42*, 71-74.

[63] H. Chen, M. M. Olmstead, D. C. Pestana, P. P. Power, *Inorg. Chem.* **1991**, *30*, 1783-1787.

[64] R. A. Jones, S. U. Koschmieder, C. M. Nunn, *Inorg. Chem.* **1988**, *27*, 4524-4526.

[65] S. C. Goel, M. Y. Chiang, D. J. Rauscher, W. E. Buhro, *J. Am. Chem. Soc.* **1993**, *115*, 160-169.

[66] C. von Haenisch, F. Weigend, R. Clerac, *Inorg. Chem.* **2008**, *47*, 1460-1464.

[67] a) L. T. Reynolds, G. Wilkinson, *Journal of Inorganic and Nuclear Chemistry* **1959**, *9*, 86-92; b) M. D. Walter, C. D. Sofield, C. H. Booth, R. A. Andersen, *Organometallics* **2009**, *28*, 2005-2019.

[68] R. A. Layfield, *Chem. Soc. Rev.* **2008**, *37*, 1098-1107.

[69] F. H. Allen, *Acta Crystallographica Section B* **2002**, *58*, 380-388.

[70] O. Kahn, *Molecular Magnetism*, VCH Publishers, New York, **1993**.

[71] F. A. Stokes, R. J. Less, J. Haywood, R. L. Melen, R. I. Thompson, A. E. H. Wheatley, D. S. Wright, A. J. Johansson, L. Kloo, *Organometallics* **2012**, *31*, 23-26.

[72] a) S. Ross, T. Weyhermueller, E. Bill, K. Wieghardt, P. Chaudhuri, *Inorg. Chem.* **2001**, *40*, 6656-6665; b) F. A. Cotton, C. A. Murillo, I. Pascual, *Inorg. Chem.* **1999**, *38*, 2746-2749; c) J. Jubb, L. F. Larkworthy, G. A. Leonard, D. C. Povey, B. J. Tucker, *J. Chem. Soc., Dalton Trans.* **1989**, 1631-1633.

[73] M. D. Fryzuk, D. B. Leznoff, S. J. Rettig, R. C. Thompson, *Inorg. Chem.* **1994**, *33*, 5528-5534.

[74] O. J. Scherer, J. Schwalb, G. Wolmershäuser, W. Kaim, R. Groß, *Angewandte Chemie* **1986**, *98*, 349-350.

[75] a) A. E. Goeta, J. A. K. Howard, A. K. Hughes, D. O'Hare, R. C. B. Copley, *J. Mater. Chem.* **2007**, *17*, 485-492; b) A. K. Hughes, V. J. Murphy, D. O'Hare, *J. Chem. Soc., Chem. Commun.* **1994**, 163-164.

[76] a) B. Y. Kimura, T. L. Brown, *J. Organometal. Chem.* **1971**, *26*, 57-67; b) F. E. Romesberg, M. P. Bernstein, J. H. Gilchrist, A. T. Harrison, D. J. Fuller, D. B. Collum, *J. Am. Chem. Soc.* **1993**, *115*, 3475-3483; c) D. B. Collum, *Acc. Chem. Res.* **1993**, *26*, 227-234; d) D. Woodruff, M. Bodensteiner, D. O. Sells, R. E. P. Winpenny, R. A. Layfield, *Dalton Trans.* **2011**, *40*, 10918-10923.

[77] J. Telser, L. A. Pardi, J. Krzystek, L.-C. Brunel, *Inorg. Chem.* **1998**, *37*, 5769-5775.

[78] J. J. Borras-Almenar, J. M. Clemente-Juan, E. Coronado, B. S. Tsukerblat, *J. Comput. Chem.* **2001**, *22*, 985-991.

[79] D. M. Low, G. Rajaraman, M. Helliwell, G. Timco, J. van Slageren, R. Sessoli, S. T. Ochsenbein, R. Bircher, C. Dobe, O. Waldmann, H.-U. Gudel, M. A. Adams, E. Ruiz, S. Alvarez, E. J. L. McInnes, *Chem. - Eur. J.* **2006**, *12*, 1385-1396.

[80] R. A. Heintz, T. F. Koetzle, R. L. Ostrander, A. L. Rheingold, K. H. Theopoid, P. Wu, *Nature (London)* **1995**, *378*, 359-362.

[81] E. Gard, A. Haaland, D. P. Novak, R. Seip, *J. Organomet. Chem.* **1975**, *88*, 181-189.

[82] M. B. Meredith, J. A. Crisp, E. D. Brady, T. P. Hanusa, G. T. Yee, M. Pink, W. W. Brennessel, V. G. Young, Jr., *Organometallics* **2008**, *27*, 5464-5473.

[83] a) J. Darkwa, J. R. Lockemeyer, P. D. W. Boyd, T. B. Rauchfuss, A. L. Rheingold, *J. Am. Chem. Soc.* **1988**, *110*, 141-149; b) T. Trinh, B. K. Teo, J. A. Ferguson, T. J. Meyer, L. F. Dahl, *J. Am. Chem. Soc.* **1977**, *99*, 408-416; cJ. E. McGrady, *J. Chem. Soc., Dalton Trans.* **1999**, 1393-1400.

[84] P. J. Alonso, J. Fornies, M. A. Garcia-Monforte, A. Martin, B. Menjon, C. Rillo, *Chem. - Eur. J.* **2002**, *8*, 4056-4065.

[85] a) E. S. Gould, *Coord. Chem. Rev.* **1994**, *135/136*, 651-684; b) M. K. Koley, S. C. Sivasubramanian, B. Varghese, P. T. Manoharan, A. P. Koley, *Inorganica Chimica Acta*

**2008**, *361*, 1485-1495; c) M. K. Koley, S. C. Sivasubramanian, B. Varghese, P. T. Manoharan, A. P. Koley, *J. Coord. Chem.* **2012**, *65*, 3623-3640.

[86] a) M. P. Marshak, D. G. Nocera, *Inorg. Chem.* **2013**, *52*, 1173-1175; b) A. Yokoyama, J. E. Han, J. Cho, M. Kubo, T. Ogura, M. A. Siegler, K. D. Karlin, W. Nam, *J. Am. Chem. Soc.* **2012**, *134*, 15269-15272.

[87] a) L.-C. Song, H.-W. Cheng, X. Chen, Q.-M. Hu, *Eur. J. Inorg. Chem.* **2004**, 3147-3153; b) L.-C. Song, H.-W. Cheng, Q.-M. Hu, *Organometallics* **2004**, *23*, 1072-1080; c) L. Y. Goh, Z. Weng, W. K. Leong, P. H. Leung, *Angew. Chem., Int. Ed.* **2001**, *40*, 3236-3239; d) L. Y. Goh, Z. Weng, W. K. Leong, P. H. Leung, *Organometallics* **2002**, *21*, 4398-4407; e) D. P. Allen, F. Bottomley, R. W. Day, A. Decken, V. Sanchez, D. A. Summers, R. C. Thompson, *Organometallics* **2001**, *20*, 1840-1848; f) F. Bottomley, D. E. Paez, L. C. Sutin, P. S. White, F. H. Koehler, R. C. Thompson, N. P. C. Westwood, *Organometallics* **1990**, *9*, 2443-2454; g) F. Bottomley, D. E. Paez, P. S. White, *J. Am. Chem. Soc.* **1981**, *103*, 5581-5582.

[88] a) G. L. Simon, L. F. Dahl, *J. Amer. Chem. Soc.* **1973**, *95*, 2175-2183; b) O. J. Scherer, S. Weigel, G. Wolmershauser, *Chem. - Eur. J.* **1998**, *4*, 1910-1916.

[89] H. Schumann, H. Benda, *Angew. Chem., Int. Ed. Engl.* **1968**, *7*, 812-813.

[90] M. Scheer, E. Leiner, P. Kramkowski, M. Schiffer, G. Baum, *Chem. - Eur. J.* **1998**, *4*, 1917-1923.

[91] S. Heinl, PhD thesis, Universität Regensburg **2014**.

[92] M. Herberhold, G. Frohmader, W. Milius, *J. Organomet. Chem.* **1996**, *522*, 185-196.

[93] O. J. Scherer, J. Braun, G. Wolmershaeuser, *Chem. Ber.* **1990**, *123*, 471-475.

[94] O. J. Scherer, G. Kemeny, G. Wolmershaeuser, *Chem. Ber.* **1995**, *128*, 1145-1148.

[95] a) H. Ogino, S. Inomata, H. Tobita, *Chem. Rev. (Washington, D. C.)* **1998**, *98*, 2093-2121; b) L. Noodleman, C. Y. Peng, D. A. Case, J. M. Mouesca, *Coord. Chem. Rev.* **1995**, *144*, 199-244.

[96] a) D. J. Brauer, S. Hietkamp, H. Sommer, O. Stelzer, G. Müller, C. Krüger, *Journal of Organometallic Chemistry* **1985**, *288*, 35-61; b) O. Stelzer, S. Hietkamp, H. Sommer, *Phosphorus Sulfur* **1983**, *18*, 279-282; c) F. Bitterer, S. Kucken, K. P. Langhans, O. Stelzer, *Z. Naturforsch., B: Chem. Sci.* **1994**, *49*, 1223-1238; d) N. A. Pushkarevsky, S. N. Konchenko, M. Zabel, M. Bodensteiner, M. Scheer, *Dalton Trans.* **2011**, *40*, 2067-2074.

[97] a) H. Lang, G. Huttner, L. Zsolnai, G. Mohr, B. Sigwarth, U. Weber, O. Orama, I. Jibril, *Journal of Organometallic Chemistry* **1986**, *304*, 157-179; b) B. E. Collins, Y. Koide, C. K. Schauer, P. S. White, *Inorg. Chem.* **1997**, *36*, 6172-6183; c) M. T. Bautista, P. S. White, C. K. Schauer, *J. Am. Chem. Soc.* **1991**, *113*, 8963-8965.

[98] L. T. J. Delbaere, L. J. Kruczynski, D. W. McBride, *Journal of the Chemical Society, Dalton Transactions* **1973**, 307-310.

[99] M. R. Churchill, J. C. Fettinger, K. H. Whitmire, *Journal of Organometallic Chemistry* **1985**, *284*, 13-23.

[100] C. M. Hay, B. F. G. Johnson, J. Lewis, P. R. Raithby, A. J. Whitton, *Journal of the Chemical Society, Dalton Transactions* **1988**, 2091-2097.

[101] S. Schulz, T. Schoop, H. W. Roesky, L. Haeming, A. Steiner, R. Herbst-Irmer, *Angew. Chem., Int. Ed. Engl.* **1995**, *34*, 919-920.

[102] C. K. F. von Haenish, C. Ueffing, M. A. Junker, A. Ecker, B. O. Kneisel, H. Schnoeckel, *Angew. Chem., Int. Ed. Engl.* **1997**, *35*, 2875-2877.

[103] a) J. L. Krinsky, M. N. Stavis, M. D. Walter, *Acta Crystallogr., Sect. E: Struct. Rep. Online* **2003**, *59*, m497-m499; b) S. Zhang, X. Zhang, Q.-s. Li, R. B. King, *Inorg. Chim. Acta* **2013**, *395*, 109-118; c) U. Koelle, F. Khouzami, B. Fuss, *Angew. Chem.* **1982**, *94*, 132; d) R. Shimogawa, T. Takao, H. Suzuki, *Organometallics* **2014**, *33*, 289-301.

[104] a) S. Weigel, G. Wolmershaeuser, O. J. Scherer, *Z. Anorg. Allg. Chem.* **1998**, *624*, 559-560; b) O. J. Scherer, S. Weigel, G. Wolmershauser, *Heteroat. Chem.* **1999**, *10*, 622-626.

[105] M. T. Bautista, P. S. White, C. K. Schauer, *J. Am. Chem. Soc.* **1994**, *116*, 2143-2144.

[106] S. Deng, Diploma thesis, University of Karlsruhe (Karlsruhe), **2002**.

[107] G. Fritz, W. Hölderich, *Zeitschrift für anorganische und allgemeine Chemie* **1976**, *422*, 104-114.

[108] Gaussian 09 Revision B.01, Gaussian, Inc., Wallingford CT, 2010

[109] a) J. P. Perdew, K. Burke, M. Ernzerhof, *Phys. Rev. Lett.* **1996**, *77*, 3865-3868; b) J. P. Perdew, K. Burke, M. Ernzerhof, *Phys. Rev. Lett.* **1997**, *78*, 1396.

[110] a) F. Weigend, M. Haser, H. Patzelt, R. Ahlrichs, *Chem. Phys. Lett.* **1998**, *294*, 143-152; b) F. Weigend, R. Ahlrichs, *Phys. Chem. Chem. Phys.* **2005**, *7*, 3297-3305.

[111] C. Adamo, V. Barone, *J. Chem. Phys.* **1999**, *110*, 6158-6170.

[112] Diamond 3.0, Crystal Impact GbR, **1997-2010**

[113] E. Keller, *Chemie in unserer Zeit* **1980**, *14*, 56-60.

[114] CrysAlisPro 1.171.36.21 or earlier versions, Agilent Technologies

[115] G. Sheldrick, *Acta Crystallographica Section A* **2008**, *64*, 112-122.

[116] A. Altomare, G. Cascarano, C. Giacovazzo, A. Guagliardi, *Journal of Applied Crystallography* **1993**, *26*, 343-350.

[117] L. Palatinus, G. Chapuis, *Journal of Applied Crystallography* **2007**, *40*, 786-790.

[118] L. Farrugia, *Journal of Applied Crystallography* **1999**, *32*, 837-838.

[119] O. V. Dolomanov, L. J. Bourhis, R. J. Gildea, J. A. K. Howard, H. Puschmann, *Journal of Applied Crystallography* **2009**, *42*, 339-341.